Food and Media

Food is everywhere in contemporary mediascapes, as witnessed by the increase in cookbooks, food magazines, television cookery shows, online blogs, recipes, news items and social media posts about food. This mediatization of food means that the media often interplays between food consumption and everyday practices, between private and political matters and between individuals, groups, and societies.

This volume argues that contemporary food studies need to pay more attention to the significance of media in relation to how we 'do' food. Understanding food media is particularly central to the diverse contemporary social and cultural practices of food where media use plays an increasingly important but also differentiated and differentiating role in both large-scale decisions and most people's everyday practices.

The contributions in this book offer critical studies of food media discourses and of media users' interpretations, negotiations and uses that construct places and spaces as well as possible identities and everyday practices of sameness or otherness that might form new, or renew old food politics.

Jonatan Leer is a Postdoc in the Department of Education, Aarhus University, Denmark.

Karen Klitgaard Povlsen is Associate Professor in the School of Culture and Communication, Media Studies, Aarhus University, Denmark.

Critical Food Studies

Series editor: Michael K. Goodman

University of Reading, UK

The study of food has seldom been more pressing or prescient. From the intensifying globalization of food, a world-wide food crisis and the continuing inequalities of its production and consumption, to food's exploding media presence, and its growing re-connections to places and people through 'alternative food movements', this series promotes critical explorations of contemporary food cultures and politics. Building on previous but disparate scholarship, its overall aims are to develop innovative and theoretical lenses and empirical material in order to contribute to – but also begin to more fully delineate – the confines and confluences of an agenda of critical food research and writing.

Of particular concern are original theoretical and empirical treatments of the materialisations of food politics, meanings and representations, the shifting political economies and ecologies of food production and consumption and the growing transgressions between alternative and corporatist food networks.

For a full list of titles in this series, please visit
https://www.routledge.com/Critical-Food-Studies/book-series/CFS

Food and Media
Practices, distinctions and heterotopias
Edited by Jonatan Leer and Karen Klitgaard Povlsen

Confronting Hunger in the USA
Searching for community empowerment and food security in food access programs
Adam M. Pine

Forthcoming

Geographies of Meat
Harvey Neo and Jody Emel

Children, Nature and Food
Organising eating in school
Mara Miele and Monica Truninger

Practising Empowerment
Wine, ethics and power in post-apartheid South Africa
Agatha Herman

Hunger and Postcolonial Writing
Muzna Rahman

Taste, Waste and the New Materiality of Food
Bethaney Turner

Food and Media

Practices, distinctions and heterotopias

Edited by
Jonatan Leer
Karen Klitgaard Povlsen

LONDON AND NEW YORK

First published 2016
by Routledge
2 Park Square, Milton Park, Abingdon, Oxon OX14 4RN

and by Routledge
711 Third Avenue, New York, NY 10017

First issued in paperback 2018

Routledge is an imprint of the Taylor & Francis Group, an informa business

British Library Cataloguing in Publication Data
A catalogue record for this book is available from the British Library

Library of Congress Cataloging-in-Publication Data
Names: Leer, Jonatan. | Leer, Jonatan, editor, author. | Povlsen, Karen Klitgaard, editor, author.
Title: Food and media : practices, distinctions and heterotopias / edited by Jonatan Leer and Karen Klitgaard Povlsen.
Description: Farnham, Surrey, UK ; Burlington, VT : Ashgate, 2016. | Series: Critical food studies | Includes bibliographical references and index. | Description based on print version record and CIP data provided by publisher; resource not viewed.
Identifiers: LCCN 2016005622 (print) | LCCN 2015043718 (ebook) | ISBN 9781472439680 (hardback)
Subjects: LCSH: Food in popular culture. | Food consumption—Social aspects. | Food preferences—Social aspects. | Television cooking shows. | Mass media—Social aspects
Classification: LCC GT2850 (print) | LCC GT2850 .L44 2016 (ebook) | DDC 394.1/2—dc23
LC record available at http://lccn.loc.gov/2016005622

ISBN 13: 978-1-138-54674-5 (pbk)
ISBN 13: 978-1-4724-3968-0 (hbk)

Typeset in Times New Roman
by Apex CoVantage, LLC

Contents

Figures and tables

Figures

Table

Contributors

Dr Vera Alexander has published extensively on issues of migration, education, identity formation, postcolonialism, diaspora, South Asian writing in English, life writing as well as Canadian literature. She is the author of *Transcultural Representations of Migration and Education in South Asian Anglophone Novels* (Wissenschaftlicher Verlag Trier, 2006). She has co-edited books on Romanticism and the Indian diaspora and has recently completed a monograph on the representation of gardens as relational spaces. She holds a position as university lecturer at University of Groningen.

Dr Susanne Eichner is Associate Professor at Aarhus University, School of Communication and Culture, Media Studies, where she is participating in a research project on the international success of Danish TV drama series. Her research focuses on reception aesthetics, media sociology, popular serial culture, video games, and the integration of an industrial and institutional focus. She is author of *Agency and Media Reception. Experiencing Video Games, Film, and Television* (Springer VS, 2014), co-author of *Die »Herr der Ringe«-Trilogie. Attraktion und Faszination eines populärkulturellen Phänomens* (UVK, 2007) and editor of *Fernsehen: Europäische Perspektiven* (UVK, 2014, together with Elizabeth Prommer) and *Transnationale Serienkultur. Theorie, Ästhetik, Narration und Rezeption neuer Fernsehserien* (Springer VS, 2013, together with Lothar Mikos and Rainer Winter).

Dr Bente Halkier is a sociologist and professor in communication at Department of Communication, Business and Information Technologies, Roskilde University, Denmark. Her empirical research focuses on food consumption in everyday life, political consumption, public consumption campaigns and the construction of consumption in media. She is the author of the book *Consumption Challenged: Food in Medialised Everyday lives* (Ashgate, 2010). She has also published on contested consumption in e.g. *Critical Public Health; Food, Culture and Society; International Journal of Consumer Studies; Journal of Consumer Culture;* and *Marketing Theory*.

Dr Joanne Hollows is a writer and researcher. Her books include *Feminism, Femininity and Popular Culture* (Manchester University Press, 2000), *Domestic Cultures* (Open University Press, 2008) and the co-authored *Food and Cultural*

Studies (Routledge, 2004). She has written numerous articles on the construction of class and gender in media food, from the Playboy era to contemporary celebrity chefs such as Jamie Oliver, Nigella Lawson, Heston Blumental and Hugh Fearnley-Whittingstall.

Katrine Meldgaard Kjær is a PhD student at the Department for the Study of Culture at the University of Southern Denmark. Her interests revolve around postcolonial theory and queer readings of contemporary culture and politics, and she is currently researching practices of inclusion and exclusion in anti-obesity and health campaigns in the US. She has talked at numerous international conferences and has written several articles on food celebrity activism and food media.

Dr Karen Klitgaard Povlsen, Associate Professor, School of Culture and Communication, Media Studies, Aarhus University. Dr Karen Klitgaard Povlsen began her many years of research on media food with her 1986 PhD thesis on fashion and food in women's magazines. She has since written on the New Nordic Cuisine and Julia Child. More recently she has conducted a study of the Danes' consumption of digital media food and she is currently working on a project on trust in organic food.

Dr Stinne Gunder Strøm Krogager, Assistant Professor, Department of Communication, Aalborg University. Stinne Krogager completed a methodologically experimental dissertation in Media Studies. Her dissertation investigates the relation between media and food in young Danes' everyday lives using novel qualitative and quantitative approaches. Currently Dr Krogager conducts research on Foodies using aesthetic processes to illustrate media use.

Dr Kathleen LeBesco, Professor and Associate Vice Dean, Marymount Manhattan College. Dr Kathleen LeBesco has published extensively on food, popular culture and fat activism. Her books include *Culinary Capital* (Berg Press, 2012), *Edible Ideologies: Representing Food and Meaning* (SUNY Press, 2008), *Revolting Bodies? The Struggle to Redefine Fat Identity* (UMass Press, 2004), *The Drag King Anthology* (Harrington Park Press, 2003) and *Bodies Out of Bounds: Fatness and Transgression* (UCalifornia Press, 2001).

Dr Jonatan Leer, PhD, postdoc, Aarhus University, has published widely on food culture, particularly on food and masculinity, in leading journals of gender and food studies, such as *Food, Culture and Society* and *NORMA: International Journal of Masculinity Studies*. His postdoc is a part of the research project on taste (www.smagforlivet.dk). His current research is on taste pedagogy and on the identity politics of the new Nordic cuisine.

Dr Peter Naccarato, Professor of English, Marymount Manhattan College. Peter Naccarato's recent scholarly work is in the area of food studies, focusing on the role of food and food practices in circulating ideologies and sustaining individual and group identities. With Kathleen LeBesco, he is co-editor of *Edible Ideologies: Representing Food and Meaning* (SUNY Press, 2008), and co-author of *Culinary Capital* (University of Georgia Press, 2012).

Dr Caroline Nyvang, Researcher at the Royal Library in Denmark. Dr Caroline Nyvang's doctoral thesis is a study of Danish printed cookbooks from 1900–70, and she has published extensively on cookbooks and recipe collections from the early modern period to the present day. In her current research project, she focuses on the cultural and social history of children's food.

Dr Fabio Parasecoli, Associate Professor and Coordinator of Food Studies at the New School in New York City. His work explores the intersections between food, media and politics. His current research focuses on food in movies, the history of Italian food and the socio-political aspects of geographical indications. Among his publications are *Food Culture in Italy* (Greenwood Publishing Group, 2004), *The Introduction to Culinary Cultures in Europe* (The Council of Europe, 2005) and *Bite Me! Food in Popular Culture* (Berg, 2008). He is general editor with Peter Scholliers of the six-volume *Cultural History of Food* (Bloomsbury, 2012).

1 Introduction

Jonatan Leer and Karen Klitgaard Povlsen

Almost all human beings enact, but also reflect on, everyday practices to do with food consumption and media use on a daily basis. It has been this way for centuries. In this anthology, we focus on the various relations and interactions between food and media: between practices of representation of food in the media and practices of interpretation of mediated food by media users.

The contribution offered by this volume lies in its presentation of a range of methodological, theoretical and empirical perspectives on food, as represented in and practised through traditional and digital media – the internet, television, campaigns, books, magazines, etc. The geographical and disciplinary diversity of the articles reflects the importance of media in various spaces of food culture, as well as the importance of food in media products and media use. Increasing interest on the part of food scholars appears worldwide in the study of food through a media lens (Appadurai, 1988; Adema, 2000; Ashley et al., 2004; Bell and Hollows, 2005; Hollows, 2008; Halkier, 2010; Johnston and Baumann, 2010; Hollows and Bell, 2011; Naccarato and LeBesco, 2012; Rousseau, 2012a, 2012b; Johnson and Goodman, 2015; Leer and Kjær, 2015). This parallels the increased interest by media scholars in the importance of food for media producers and consumers in the past ten years (Bonner, 2005; Reilly, 2006; Miller, 2007; Bonner, 2009; Krogager et al., 2015; Rittenhofer and Povlsen, 2015; Thorsø et al., 2016).

In this volume we are not advocating the founding of a new concept or research tradition. However, we see the relation between food and media as interactive, and extremely important, as media pervade all spheres and all chains of contemporary food-ways, from certified labels to television chefs and blogs and recipes on the internet. So our overall aim with this volume is to show that contemporary food studies need to pay more attention to the significance of media in relation to how we 'do' food. Media need to be addressed in food studies, not only when they are at the centre of a project (for instance when studying gender and taste ideals in food television), but also when they are less obvious and less central (for instance an investigation of children's food preferences could gain immensely from understanding how media play a role in children's relations to food) – as demonstrated by Krogager and Eichner in this volume.

This volume offers ways to initiate and develop the media dimension of food cultures – both methodologically and theoretically – in its numerous arenas. The contributors move from single-media studies such as cookbooks or TV shows to cross-media studies of the uses of cookbooks, magazines, TV and internet. The volume thus shows the development from representation to reception or use; from studies that concentrate on the producer and texts to studies that focus on the receiver or user. Cultural studies, food studies and media studies are thus combined.

Though the chapters are diverse in their historical and geographical scope and in their methodological and theoretical perspective, all the contributions in this volume are unified, first and foremost, by their focus on articulations of sameness and diversity, difference and 'otherness,' in single as well as cross-media content and among media users. A second common factor is sensitivity towards the cultural contexts in which media are embedded. Although we see a tendency towards globalisation of media contents, we also see a strong tendency towards local contents and uses. Media formats and platforms are becoming more alike worldwide, yet at the same time more diverse in relation to use and in relation also to contents that are co-produced by users. Media and food may travel across cultures through export, in which case they may stay the same, or they may be transformed or adapted. Thus on the one hand we see the politics of media food as hegemonic, globalised and commercialised; but on the other, we are aware of the constructions of local diversities and cultural capital that are articulated in media food and in the uses and negotiations of the mediated representations.

These two common features are reflected in the chapters in this volume by the shared methodological understanding of food and media as 'sayings,' as well as 'doings.' On both macro and micro level, food-ways and identity are considered cultural practices – both done and as said – that are socially constructed, and that should therefore be understood in relation to the dynamics of their historical, social and cultural context.

Furthermore, three ideas run through the contributions and structure connections between the chapters. Firstly, all contributions understand media and food as practice, and accordingly discuss the composite relationship between media and food practices. An important key to comprehending this intertwinement is the second shared theme, that of distinction and taste. In the contributions of the book, both food practices and the uses and interactions with media tend also to be practices of taste. Hence, the social and mediated spaces in which both sets of practices are enacted have particular potential to express belongingness, diversity and a growing range of distinctions. The volume thus goes beyond Bourdieu in its attempt to rearticulate his central term of distinction. Throughout the volume, we operate with a post-Bourdieusian perspective as we understand distinctions as social *and* individual constructions in all spheres of life. Food distinctions and media distinctions intersect with other lifestyle distinctions. As taste practices demarcate boundaries and openings both locally and globally, a spatial approach is helpful in analysing the spaces of food and media. So, thirdly, Foucault's reflections on counter-spaces – named heterotopias – and their relations to hegemonic spaces are suggested as an instrument to grasp the cultural significations and paradoxes of the spaces of distinction created by the food and media constellations. It is certainly in the operationalisation of the concept of heterotopia for understanding the relationship between practice, distinction, food and media that the original contribution of this volume lies.

In this introduction, we develop these three ideas and we argue that the ideas of practice, distinction and heterotopia offer a productive way of thinking about and analysing the entanglements of food and media. Our goal is not to offer a one-size-fits-all solution for how to study food and media. Rather, the volume proposes a multitude of perspectives on the field, and a series of tools to investigate food and media within this framework. Our introduction to the chapters and their approaches to food and media will serve to illustrate this last point.

Media and food practices

In and outside media studies, an intense debate on mediatization and mediation (Couldry and Hepp, 2013; Jensen, 2013; Krotz and Hepp, 2014) has developed during the last ten years, with new considerations on the interplay between media, culture and society (Schulz, 2004; Krotz, 2007; Hjarvard, 2008; Livingstone, 2009; Lundby, 2009; Krotz and Hepp, 2014; Hepp et al., 2015) – not least in understanding the increasing relevance of media to various fields of society. Mediated communication or *mediation* is the more general term, denoting communication or representation processes, such as food as content in various media such as TV, the internet, magazines, etc. (Livingstone, 2009; Couldry, 2012; Livingstone and Lunt, 2014). *Mediatization,* on the other hand, is the conveying of historical transformations in social and cultural environments through communication and media, such as the growing importance of media institutions in society (Hjarvard, 2009; Hepp et al., 2015). The debate has sensitised media research for studies that empirically demonstrate how media institutions, media products, media texts and uses of digital media or mass-media consumption have become seamlessly interwoven in all aspects of everyday consumption and practices, including food (Jensen, 2013).

Media practices play a central differentiated and differentiating role in people's everyday practices in relation to food, both in and outside the home. The interplay between what we eat and which media we use is complex, and it is best understood when we regard the two fields of consumption as an interrelating influence (Simmel, 1972). In this volume our understanding of mediatization lies on a par with Friedrich Krotz's focus on the transformative potential of media tools as used by ordinary people to construct their social and cultural world (Krotz, 2009, pp. 24–25). Thus every historical period realises the interrelation between media and cultural fields such as food in specific ways that are to be explored empirically. Vera Alexander's chapter offers a nineteenth-century example of how letters, drawings and books become tools for understanding new food items and constructing new food-ways in daily life. Jonatan Leer does the same in his chapter on how contemporary TV chefs construct taste regimes and masculinity on TV. As a whole the volume attempts to present specific cases on how media play an increasing role in culture and society:

> It seems that we have moved from a social analysis in which the mass media comprise one among many influential but independent institutions whose

> relations with the media can be usefully analysed to a social analysis in which everything is mediated, the consequence being that all influential institutions in society have themselves been transformed, reconstituted, by contemporary processes of mediation.
>
> (Livingstone, 2009, p. 2)

The concept of mediatization is such a move – one that can be researched in relation to empirical cases of social constructions of media, identity and culture in people's daily lives. This has created a renewed interest in media as relays for both individual and societal food practices, focusing on 'normal' and routinised media practices embedded in the context of everyday life – which is why there is also a renewed interest in practice theory in media studies (Couldry, 2004; Halkier, 2010).

Practice theory lies in line with the social-constructivist view of human beings' transformation and construction of the world they live in and by. Social structure and human agency are understood as a dynamic dialectic (Giddens, 1984). Bourdieu's *Outline of a Theory of Practice* (1977) demonstrated how practice theory might be applied to empirical data and cases. Reckwitz's definition of practice makes it evident that it is highly relevant also for media and cultural studies:

> A 'practice' (Praktik) is a routinised type of behaviour which consists of several elements, interconnected to one other: forms of bodily activities, form of mental activities, 'things' and their use. . . .
>
> A practice – a way of cooking, of consuming, of working, of investigating, of taking care of oneself of others, etc.
>
> (Reckwitz, 2002, pp. 249–250)

All these activities are interwoven in media representations or uses of media. An entertaining TV show might inform viewers of new food items, or ways of cooking. Some viewers seek more information on the internet, or recipes, and look at labels or texts on food wrappings before they buy it, cook it and consume it. Food consumption is more often than not embedded in cross-media uses, from print to electronic and digital media of all sorts. Just as most media use and most food consumption is individual and social at the same time, the individual subject is a crossing-point for food and media practices – but in a social context of work, family ties, and living circumstances.

While practice theory, with a few exceptions (Couldry, 2004, 2012; Halkier, 2010; Hepp, 2013), has been blind to media as a relay to other practices, media researchers are beginning to see the possibilities in practice theory for media representations, as well as media reception, because of the intense mediatization of late modern societies. Couldry (2004), Hepp et al. (2015), and Krotz and Hepp (2014) were the first to urge media sociologists to reorient their research towards practice theory. The orientation towards practice theory in this volume is focused on practice in relation to discourse and interpretation, or sayings as doings (in part I); and on practice in relation to the production and use of media and media content, or doings as sayings (in part II) (Couldry, 2010, p. 2). Both fields are contained in the space of practices that organise our everyday lives, framed by conventions, habits and material, and technological possibilities and restrictions. In both fields, sayings and doings are considered actions, and in both fields, hierarchies, conventions and distinctions make some practices more mainstream than others, as Couldry argues in his debate with the anthropologist Hobart (Couldry and Hobart, 2010). In both fields, other practices than media and foodways exist, but in this volume we only consider media-related food representations and practices.

Already in 2004 Couldry, inspired by Bourdieu (1977), Schatzki (1996, 2002) and Reckwitz (2002) and recent empirical sociology, suggested that a new paradigm was emerging in media studies, namely practice: 'it treats media as the open set of practices relating to, or oriented around, media' (2004, p. 4). This would free media studies from the insoluble problem of how to prove media effects, and at the same time make *non*-media-centric investigations possible, as in this case the intersectional spaces of practices between media and food. What people actually do with media in relation to food in differing contexts and situations – for instance, TV cooking shows – needs further investigation. The role of media cannot be isolated in culture; that it is often articulated by people in different ways (Bird, 2003, pp. 2–3) suggests that media play differing roles in people's everyday lives.

Couldry's suggestion has only recently begun to resonate in connection with the debate on mediatization, which underlines that more empirical research is needed on how media in media-saturated societies are actually used, and to which means. Media practices need to be researched in an everyday context, at home or in institutions, and not least in relation to food. In this volume, we concentrate on the domestic field of food media in leisure and home, and on the domestication of new media-related food-ways (Silverstone, 2005; Berker et al., 2006; Hartmann and Hepp, 2010) in contexts from single-media representations (i.e. print, TV) to cross-media uses (for instance with smartphones, which incorporate most other media). We offer empirical cases not of 'the media,' but of selected media and cross-media patterns (Couldry, 2009), where we look at mediatization as embedded communication (Jensen, 2013, p. 216): embedded in the routines and habits of everyday life, but embedded also in material and technological and societal structures (Couldry and Hepp, 2013, p. 191). In many ways, this parallels the ways in which food and food-ways are embedded in everyday life, as well as in larger societal structures such as diverse media representations.

Schatzki defines the field of the 'practice approach' as 'all analyses that 1) develop an account of practices. . . . 2) treat the field of practices or some subdomain thereof' (Schatzki, 2001, p. 11). In this volume, we use the term 'field' in a broad sense, not limited to Bourdieu's definition. As a synonym, we use the term 'spaces of practice,' because this is relevant in relation to Foucault's term 'heterotopia,' which will be developed later in this introduction.

Productively applying practice theory is Halkier, in her chapter on everyday mothering and two genres of food media: Danish women's magazines, and public campaigns intended to influence Pakistani women living in Denmark towards a healthier diet. The practice-theoretical approach sees media discourses as resources that are being domesticated in differing ways in the two groups, which nevertheless share a number of ways of doing mothering. The domestication perspective proves useful to an understanding of media in everyday life that is not media-centric (Pink and Mackley, 2013, p. 681) but focuses on the uses, negotiations and effects of a media, a genre or a theme in media discourses. The multiplicity of different media discourses – across different genres and platforms – means that the constellations of actual food and media practice are enacted and performed. This makes it meaningful to compare two different media genres, two different groups of mothers and many different media practices relating to food and mothering. What is shared is that offering good food to children and families is considered to be an important part of mothering in both groups, but the ways in which this ideal is enacted vary according to cultural traditions, media use and daily routines. The normative expectations are important, and different. While the ideal among Danish magazine-readers is cooking from scratch, the ideal among Pakistani women in Denmark is traditional Pakistani food. But both groups bend these ideals – and partly because of the help offered by media discourses – in their everyday routines, in ways that only become clear when they are studied as complex practices of mothering related to media uses and food consumption.

Another perspective is offered by Povlsen in her article on how Danes talk about and demonstrate their media routines related to food. Povlsen's chapter looks at two samples of qualitative interviews, undertaken as a follow-up to two Danish surveys on digital media use and trust in organic food and labels. The most impressive result was the importance of a search engine such as Google. All but one respondent spontaneously answered that the most important media to find recipes was 'to google.' When discussed in more detail, however, it became evident that the respondents, except one, were not reflecting upon what they actually do when they google, or on why they choose the websites they do.

Povlsen's chapter demonstrates the difficulty of using traditional sociological segmentations or groupings. Her material shows that the enacting of and reflection on daily routines and practices concerning media and food are at one and the same time both deeply personal (and dependent on specific family patterns and educational backgrounds), yet first and foremost conditional on the individual's health or interest in health and fitness. Of course having small children, or commuting for hours between home and job, or living in a rural area without supermarkets will influence not only food routines, but also media routines and the relations between the two. Against traditional expectations, the data demonstrates that those with chronic diseases and the elderly are not only the heaviest, but also the more competent media users, especially in relation to digital media use. The data also demonstrates that the much-discussed category of 'foodies' (Johnston and Baumann, 2010) is a difficult one to use, because people's sayings are often not congruent with their doings, and because their media uses also often point to the fact that mainstream media use sometimes parallels mainstream food tastes, even though what is said in the interviews constructs foodie discourses. The article therefore incorporates Foucault's concept of *heterotopia* to understand media and food practices as spaces that allow for sameness and diversity in both fields.

Eichner again takes another approach towards practice theory in media consumption in media-saturated societies like Germany. She focuses on where German children aged between six and twelve encounter food as content in the media, and also on their actual media uses and preferences. Taking the statistical top three or top five media texts preferred by German children, Eichner constructs an analytical sample of diverse media texts from diverse genres and platforms. The statistics on the popularity of texts hint at the actual media practices of German children, but as Eichner did not have the opportunity to observe or interview children, she is dependent on a quantitative content analysis, followed by close readings of texts as examples of reception-aesthetic analyses, using the theory of reception aesthetics as advanced by H. G. Jauss.

Eichner's research shows us that children are not a homogeneous audience. Certain particularities of media use regarding age, gender, nationality and regionality become evident both in the statistical findings and in the content analyses. What the chapter does offer, however, is a mapping of food in the media that is actually used and preferred by the children. In Germany, TV is still the most important media for children, while online media activities play a minor role. Food representations are often embedded in fictional formats, and they often focus on sweets and beverages. The perspectives and potentials for lifestyle and distinction strategies seem obvious, but are also dependent on user or audience studies that might take the orientation towards actual media and food practices a step further in qualitative observations and interviews.

Such a step is taken by Krogager in her article on how Danish adolescents around the age of sixteen actually co-produce food campaigns and other food content for traditional media formats, such as advertising or health campaigns. The study performed was inspired by experimental methods, such as media productions by focus groups conducted during school hours, but outside the traditional classroom setting. Groups of boys and groups of girls were encouraged to produce a health campaign with a slogan and pictures, or to compose an advertisement using the same basic food, such as a chicken or a burger. The groups had limited time and were observed by researchers. After the experiment, a lengthy focus-group interview was carried out with a loose interview guide, making room for the interactions and ideas from the group to unfold.

The study thus demonstrates the doings and sayings in a group of adolescents exhibiting their media and media-literate competencies in their adaptations of traditional media genres and campaigns. An example is how the McDonalds slogan, 'I'm lovin' it,' is reversed to the health slogan of 'I'm hatin' it.' The media productions showed a deep understanding of media mechanisms and a broad knowledge of healthy/unhealthy foodstuffs to a degree that the interviews alone could never have done. The contribution is thus also demonstrating the limitations of direct questions when interviewing this age group, as well as the advantages of methods inspired by practice theory.

Distinction and taste

Uses of or reception of print, electronic and digital media as well as diverse food preferences often demonstrate traditional patterns of class, education, gender and ethnicity à la Bourdieu. They are also often seen as social performances or enactments of exclusion from and inclusion in social groupings and taste regimes, inspired by Simmel (2008) and others. This volume contributes to contemporary debates around food and distinction from a post-Bourdieusian perspective. According to Bourdieu, rather than the expression of individual preference or personalised patterns of consumption patterns, taste is 'the propensity and capacity to appropriate (materially and symbolically) a given class of classified objects and practices,' and as such 'the generative formula of a lifestyle, a unitary set of distinctive preferences which express the same expressive intension in a specific logic of each other of the symbolic sub-spaces' (Bourdieu, 1984, p. 173). For Bourdieu, an analysis of class and differences in food practices functions to illustrate that taste is an important expression of belongingness and distinction through everyday practices, as 'social identity is defined and affirmed by difference.' Media practice is another such subfield in which daily practices express social position.

Bourdieu's ideas have had a huge impact on studies of tastes, but they have also been heavily debated and challenged. Especially the limited possibility for individual agency in Bourdieu's account has been problematised. Several scholars have stressed that Bourdieu's deterministic perspective did not correspond to (post)modern/post-1968 consumer society, in which lifestyle was no longer dictated by traditional structures (class, religion, geography), and that these societies could be described as post-traditional (Giddens, 1991) or fluid (Bauman, 1988, 2000). In these societies, each subject could balance his or her identity on a 'day-to-day social behaviour' (Giddens, 1991). Furthermore, the new lifestyles and subcultures proposed in late capitalist societies gave everyone (at least in the growing middle classes) the 'choice' to choose or zap across new, ad hoc communities or 'tribes' (Maffesoli, 1996). The field of food offers plenty of examples of such tribal identities: Paleo, vegan … Sociologist Alan Warde opposes these two positions on taste, on the one hand a perspective that emphasises distinction and social reproduction (Bourdieu), and on the other, one that emphasises freedom, choice and mobility (Baumann). Warde argues that in order to better our understanding of taste and food practices as individual and collective, we have to reconcile the two perspectives: 'No two people will exhibit identical behaviour. However, and this is an empirical matter, there may remain considerable similarities between some individuals, and systematic differences between one group and another' (Warde, 1997, p. 3). So although Warde nuances Bourdieu's perspective, he concludes that 'tastes are still collectively shared to a very significant extent' (Warde, 1997, p. 3).

The idea of omnivore consumption has been another way of reconciling the two positions. This thesis suggests that we currently see that the distinctions between high and low culture are no longer straightforward. Particularly in the middle class, we see new omnivore consumers shopping around in the cultural repertoires (Johnston and Baumann, 2010). Although such patterns of consumption are antithetical to snobbishness, this 'does not signify that the omnivore likes everything indiscriminately, rather it signifies an openness to appreciating everything' and 'its emergence may suggest the formulation of new rules governing symbolic boundaries' (Peterson and Kern, 1996, p. 904). So although the omnivorous tendency has democratic potential, it also creates new patterns of distinctions in relation to food:

> The omnivorous age does not usher in relativistic cultural paradise where "anything goes" and all foods are made legitimate. Instead, boundaries between legitimate and illegitimate culture are redrawn in new, complex ways that balance the need for distinction with the competing ideology of democratic equality and cultural populism.
>
> (Johnston and Baumann, 2007, p. 179)

The aim of this volume is to bring nuance and further discussion to these new dynamics of distinctions in relation to food and media through the empirical cases presented in the single chapters. The volume's perspective on distinctions thus shares many perspectives with the concept of 'culinary capital' in seeking to understand 'why certain food and food related practices connote, and by extension, confer status and power on those who know about and enjoy them' and 'how specific cultural spaces shape individuals in dramatic ways, rewarding those who succeed with culinary capital and denying it to those who fail' (Naccarato and LeBesco, 2012, p. 3). This concept – explicitly inspired by Bourdieu – offers a way of using his central idea of distinction, but in social contexts that are more fluid, composite and omnivore than that described in *Distinction*, and paying particular attention to the importance of space and the spatial diversity of mainstream and subcultural practices of food and media, while also bearing in mind the exclusiveness of such spaces and practices. The central point that flows through many of the chapters in this volume is the idea of ongoing and never-ending negotiations of ranges of taste and distinctions. A food, media or taste practice is not legitimate/ or illegitimate per se, but depending on the status and social identity of the person or social group performing the taste practice, it can be, become or be negotiated as an illegitimate practice.

The chapter by Nyvang illustrates this point by analysing how adaptations of 'foreign' food cultures can be used to create culinary proto- and antitypes, and also to engage in a debate about taste ideals and gender distinction. More specifically, Nyvang's chapter analyses how perceptions of American and French food were used as models in a postwar cookbook published in Denmark. The analysis identifies two groups partaking in the discussion of the direction of Danish food culture on the brink of a new era. On the one hand, there is the female consultant inspired by the innovative kitchen technologies and rapid recipes from the US that made it possible for women to undertake a professional career while catering to the family and fulfilling the housewife role. On the other hand, there are the gourmands, a group of male artists and engaged *amateurs de cuisine* who embrace the individual sense of taste and the hedonistic ideals of the masculine, French *gourmand* tradition *à la* Brillat-Savarin. Nyvang examines the clash between the two positions in the public debate, and shows how they echo a traditional gender opposition in Western food culture between 'feminine' cooking as a care-for-others project and 'masculine' cooking as a care-for-self project (Lupton, 1996, p. 134). The article shows how gender and culinary capital are connected, and how specific kinds of culinary capital are accessible to men and women. Furthermore, Nyvang's mapping of these competing discourses adds an important perspective to the debate on food, media and distinction: namely, that different understandings of culinary capital – and accordingly different political agendas – coexist in a struggle for hegemony, and that these struggling understandings play an important role in these food fights.

Alexander also places gender, taste and distinctions in the foreground in her historical chapter on mediated food in the writings and drawings of two sisters, Catharine Parr Traill (1802–99) and Susanne Moodie (1803–85), who in the 1830s emigrated from civilised England to the wild outskirts of Canada. These two pioneers published personal narratives, telling other English pioneers what to expect and how to prepare for the Canadian bush. So food is something to write home about, but also a creative challenge of wilderness, not least for a woman. Alexander's close reading of these texts shows how the preparation of food intersects with the invention of a new feminine selfhood that combines handicraft with femininity – with results such as dandelion coffee. Food was an area of control in the wilderness, a necessity for survival. Canadian foods such as maple syrup, transformed and represented in English terms, gave the women comfort and aesthetic pleasure. The production of food and food writing thus connect past and future, Englishness and Canadian-ness. Food establishes a relational space or heterotopia on many levels, not least identity and gender roles. The chapter thus demonstrates the productivity of Foucault's term heterotopia for textual close readings of older texts.

Negotiating hegemonies and constructing heterotopias

As the post-Bourdieusian perspective on distinction makes evident, media-food 'texts' offer recipes for 'a better life' or a 'promise of happiness' (Ahmed, 2010) in a world in which norms, practices and lifestyles are fluid and negotiable, while remaining to a large degree structured by new moral regimes and ideas of cultural legitimacy as well as being shaped by material and economic constraints.

However, in the heterogeneous and multi-fractioned mediated food fields or spaces, there exists no consensus on what 'the good life' actually is, or on how to enact it in relation to food consumption, cooking and social eating. There is, however, a consensus that the good life exists. The various practices and combinations of food and media could be seen as a battlefield of competing identity discourses, discourses that seek legitimate ground to claim hegemony by or through food. But if we consider food representations in the media to be a 'space' in which deviant positions are idealised or condemned yet articulated, food media are conceptualised not only as a hegemonic force, but also as diverse platforms for the emergence of counter-positions, and therefore of possible sites for constructions of otherness.

In order to articulate this ambiguous connection, we propose Foucault's concept of heterotopia (Foucault, 1997). This offers us possibilities for conceptualising both hegemonic and counter-hegemonic dimensions of the spaces offered by food and media. Foucault defines the heterotopia in opposition to the utopia as:

> something like counter-sites, a kind of effectively enacted utopia in which the real sites, all the other real sites that can be found within the culture, are simultaneously represented, contested, and inverted. Places of this kind are outside of all places, even though it may be possible to indicate their location in reality.
>
> (Foucault, 1997, pp. 333–334)

A common feature of heterotopias, according to Foucault, is that through the specificity of their spatial and temporal quality they form counterpoints to – but also connections with – the spaces of the surrounding culture. This rather sketchy definition is not made more precise by Foucault's subsequent development of a 'heterology,' a descriptive list of principles of the heterotopias, nor by the very heterogeneous list of enumerated examples of heterotopias presented in the text. According to Foucault, heterotopias can be cemeteries and ships, museums and hammams, libraries and holiday villages. What all these places have in common according to Foucault is that, through the specificity of their spatial and temporal quality, they form counterpoints to – yet at the same time they are also somehow connected to – the spaces of the surrounding culture. Unlike utopias, which are idealised counter-spaces, heterotopias are actually localisable.

This concept of the heterotopia has been popular within the cultural and social sciences (Johnson, 2013, e.g. the list of thirty-six examples of the use of the concept, pp. 796–797), to such a degree that the already elusive idea in Foucault's sketch has become emptied of sense as its many uses are said to be 'contradictory,' 'opposed to each other,' and occasionally even 'completely incomparable' (Ritter and Knaller-Vlay, 1998, p. 14), as if everything – and nothing – could be described as a heterotopia.

The loose definition limits the concept, and the fragmentary and inconsistent character of Foucault's text has certainly given rise to very free interpretations and applications. Johnson (2013) warns in an elaborated article against the dangers of simplistic application, especially in studies emphasising the revolutionary or subversive counter-dimension of heterotopias. A heterotopia does not designate absolute difference from society or from hegemonic spaces. Rather, Johnson argues that heterotopias 'are distinct emplacements that are "embedded" in all cultures and mirror, distort and react to the remaining space' (Johnson, 2013, p. 794). It is exactly this point that we find stimulating in relation to food and media entanglements; and when used in this way, the idea of heterotopia – and the many openings offered by its lack of fixation – has proved very useful in allowing us to avoid simplifying our understanding of the impact of food media. Foucault's concept is a flexible tool that can be used to demonstrate how food media texts are in fact sites for diverse and opposing hegemonic discourses, creating counter-discourses and counter-spaces. It also allows us to demonstrate how the users of food media accept, resist and negotiate the codes, norms and politics of media-food discourses in their everyday adaptations. Looking at food and media through the lens of heterotopia casts light on the politics of food media and on their possible relations to practices of many sorts. This is a vision that can take the cultural specificity of the hegemony and heterotopia of food and media into account: a vision that pluralises the very idea of hegemony and heterotopia, in order to see and talk about the hegemonies and heterotopias of food. This perspective runs through the volume, but the many distinct empirical settings also demonstrate that the relationship between hegemony and counter-hegemony is not a straightforward one, nor can we make universal generalisations about it. Throughout the book, however, the chapters again and again demonstrate the productivity of the term heterotopia for understanding the complex intersectionalities of the single empirical cases. As we have already seen in the case of Alexander's chapter, heterotopia works fruitfully as an analytical and theoretical tool – or even 'can opener' – to add depth to the analysis of the intersection of gender, migration and food writing. In the following, we'll demonstrate how the concept of heterotopia also develops the empirical analyses in other chapters.

The development of the mediated food scene has created a new 'food nobility' – celebrity chefs and amateur chefs, doctors and health professionals, diet and environmental experts – who are the promoters of various recommendations on how to live better through food: from the Paleo diet to ethical or environmental consumption, from fast food to slow food, from Rachael Ray's *30-Minute Meals* to Hugh Fearnley-Whittingstall's self-sufficient agricultural heaven in the *River Cottage* shows. This 'celebrisation' of food culture opens a new series of paradoxes about food practices and rearticulates ideas of exclusiveness and accessibility, as well as ideas of responsibilisation and empowerment in relation to food (Johnson and Goodman, 2015). This trend thus rearticulates debates of distinction mentioned above in relation to both taste and food practices, as well as in relation to debates of societal or individual agency and guilt. A major concern could be that the celebrity food activism individualises problems by a kind of neoliberal 'You-can-if-you-want-to'-logic (Leer, in press) that does not articulate the structural issues concerning food. This is also the central point in Hollows's chapter in this volume, in which she demonstrates how in several new TV shows featuring British male celebrity chefs and their campaigning activities, the male celebrities' enterprises are articulated as counter-hegemonic. These food 'celanthropists' (Rojek, 2014) are portrayed as men who care and 'do something' in relation to a bigger food issue that the state has failed to deal with, and they offer alternative solutions to save the nation. At the same time, their ideas of class and gender are extremely hegemonic, as Hollows demonstrates. So as in Foucault's conceptualisation of the heterotopia, hegemony and counter-hegemony are interrelated in various ways and should not be considered as absolute differences. Hollows's contribution also underlines the intersection between class and gender, as the campaigning culinary documentaries create a middle-class gaze that works to highlight 'bad' feeding practices and associates them with working-class women.

Kjær's chapter, on Michelle Obama's food activism, also addresses the ambivalence of the space of celebrity identity in relation to race and gender, novelty and tradition, public and private spheres. Kjær concludes that Michelle Obama 'is actually negotiating a space for an African-American First Lady to be eligible to be considered the above-discussed site for traditional femininity that represents American womanhood.' Here, the identity negotiations articulated through engagement with media and food are clearly spatialised, and it is made evident how Michelle Obama has to navigate between a series of hegemonic and heterotopian tropes and spaces which are both actual spaces (such as the White House) and virtual spaces of the media sphere; it is in this spatial navigation that her identity negotiation takes place.

Parasecoli's article, on former basketball player Charles Barkley's engagement in weight-loss commercials, exposes a similar negotiation of gender, race, body ideals and 'celebrityness' in relation to food and media. A central feature of this negotiation is also the spatial opposition between the basketball arena (a site for hegemonic masculinity) and the virtual space of the weight-loss commercials (a traditionally deeply emasculated space). However, in the commercials and through slogans like 'I have won like a man as long as I have been a man. Now I am losing like one,' there seems to be an attempt to make the heterotopian scene for weight-loss hegemonic. So once again, we witness a complex interplay between heterotopian and hegemonic spaces in the fields of food and media. However, as in Hollows's chapter, hegemony seems to pervade and gain the upper hand as Parasecoli concludes that 'media participates in the production and naturalisation of cultural bias, social dynamics, and power hierarchies by turning the norms, negotiations, and the tensions that underpin them into entertainment and marketing.' Here, Parasecoli also underlines a fundamental point about food and media, one that is echoed in several of the other chapters: namely, that media and food are still among our strongest cultural markers and markets, and that we must not be blind to the commercial and political systems in which mediated food, like its recipes and its promises, are always embedded. However, as the examples of food celanthropy reveal, the relationship between market and markers, between hegemony and counter-hegemony, is not straightforward. The boundaries around the spheres are fluid.

Leer sets out to describe a new masculine tendency in TV cooking shows. He shows how male TV chefs appear to use cooking as a way of escaping the post-traditional imperative for gender negotiation. Cooking becomes a way of performing male bonding in a zone free from 'feminine' food morality. This is done through an analysis of the popular Danish show *Spise med Price*. In each episode of the show, two brothers cook together in an isolated summerhouse. Leer argues that the summerhouse could be read as a homosocial heterotopia, created through the use of various transgressive and out-of-the-ordinary culinary practices and discourses. This reading also develops Foucault's heterotopia concept by operationalising his idea of the heterochronic character of the heterotopia. This is done by demonstrating how nostalgia and 'dated' discourses are developed in this space and used to masculinise the spaces and the cooking of the show. At the same time, this is done in a very ironic manner, which also renders this gendering ambivalent.

Ideas of heterotopias are also articulated in the volume by analyses of actual uses and of perceptions of food in the media. Halkier argues that both her empirical case-studies on how food and media are used to negotiate motherhood have examples that could be interpreted as heterotopian performances, in which the dominant discourses on appropriate motherhood are bent in everyday cooking. In this sense the domestic kitchen becomes a place invaded by hegemonic media discourses, and although these to a large degree shape ideals of cooking and motherhood, there is also some room for counter-hegemonic practices and resistance to hegemony in this same space.

Povlsen's contribution on cross-media use in relation to food also argues that the adaptations of media-food text are extremely complex and ambivalent practices. The analysis of her interviews with a very varied range of Danes underlines that the food omnivore is also for the most part a media omnivore, but at the same time, 'the majority is no homogeneous mass.' People tend to use media for various purposes and participate in a variety of taste regimes. One common feature, however, is that all participants use food and media to create heterotopias that in various ways are integrated into everyday routines. Povlsen concludes that in the fluid and multi-fractioned world of today, Foucault's idea of the heterotopia can provide a new perspective to rethink Bourdieu and make his essential idea of distinction relevant in this era of fluid cultures.

We are happy that the authors of *Culinary Capital,* Peter Naccarato and Katie LeBesco, agreed to write an epilogue for this volume in which they also venture some reflections on the future of food and media. They predict a future in which media producers continually adapt their programming in order to attract and maintain the interest of the viewing public. And for media consumers, it will be one in which they exercise increasing influence over the types of media produced and the extent to which that programming functions hegemonically or counter-hegemonically.

With this in mind, we anticipate that there is lots of work ahead for food (and media) studies in order to understand the transformations and expansions of the field. New research designs and methodologies are needed. New areas and connections between food and media need to be approached and explored. Hopefully, this volume will inspire its readers to engage in such endeavours.

References

Adema, P., 2000. Vicarious consumption: Food, television and the ambiguity of modernity. *Journal of American & Comparative Cultures*, 23(3), pp. 113–124.

Ahmed, S., 2010. *The Promise of Happiness*. Durham: Duke University Press.

Appadurai, A., 1988. How to make a national cuisine: Cookbooks in contemporary India. *Comparative Studies in Society and History*, 30(1), pp. 3–24.
Ashley, B., J. Hollows, S. Jones, and B. Taylor, 2004. *Food and Cultural Studies*. London: Routledge.
Bauman, Z., 1988. *Freedom*. London: Open University Press.
Bauman, Z., 2000. *Liquid Modernity*. Oxford: Polity Press.
Bell, D., and J. Hollows, eds., 2005. *Ordinary Lifestyles*. Maidenhead: Open University Press.
Berker, T., M. Hartmann, Y. Punie, and K. Ward, eds., 2006. *Domestication of Media and Technology*. Basingstoke: Open University Press.
Bird, E.E., 2003. *The Audience in Everyday Life*. London: Routledge.
Bonner, F., 2005. 'Whose lifestyle is it anyway?' In: D. Bell and J. Hollows eds., *Ordinary Lifestyles*. Maidenhead, Berkshire: Open University Press, pp. 35–46.
Bonner, F., 2009. Early multi-platform in television. *Media History*, 15(3), pp. 345–358.
Bourdieu, P., 1977. *Outline of a Theory of Practice*. Cambridge: Cambridge University Press.
Bourdieu, P., 1984. *Distinction*. London: Routledge.
Couldry, N., 2004. Theorising media as practise. *Social Semiotics*, 14, pp. 115–132.
Couldry, N., 2009. Does 'the media' have a future?. *European Journal of Communication*, 24(4), pp. 437–449.
Couldry, N., 2010. 'Theorizing media as practice.' In: B. Bräuchler and J. Postill eds., *Theorising Media and Practice*. New York and Oxford: Berghahn Books, pp. 35–54.
Couldry, N., 2012. *Media, Society, World: Social Theory and Digital Media Practice*. Cambridge: Polity.
Couldry, N., and A. Hepp, 2013. Conceptualizing mediatization: Contexts, traditions, arguments. *Communication Theory*, 23, pp. 191–202.
Couldry, N., and M. Hobart, 2010. 'Media as practice: A brief exchange.' In: B. Bräuchler and J. Postill eds., *Theorising Media and Practice*. New York and Oxford: Berghahn Books, pp. 77–82.
Foucault, M., 1997 (1967). 'Of other spaces: Utopias and heterotopias.' In: N. Leach ed., *Rethinking Architecture: A Reader in Cultural Theory*. London and New York: Routledge, pp. 330–339.
Giddens, A., 1984. *The Constitution of Society: Outline of the Theory of Structuration*. Cambridge: Polity Press.
Giddens, A., 1991. *The Consequences of Modernity*. Stanford: Stanford University Press.
Halkier, B., 2010. *Consumption Challenged*. Burlington: Ashgate.
Hartmann, M., and A. Hepp, eds., 2010. *Die Mediatisierung der Alltagswelt*. Wiesbaden: VS Verlag.
Hepp, A., 2013. *Cultures of Mediatization*. Cambridge: Polity.
Hepp, A., S. Hjarvard and K. Lundby, 2015. Mediatization: Theorizing the interplay between media, culture and society, *Media, Culture & Society*, 37(2), pp. 1–11.
Hjarvard, S., 2008. *En verden af medier*. Frederiksberg: Samfundslitteratur.
Hjarvard, S., 2009. Samfundets medialisering. *Nordicom-Information*, 31(1–2), pp. 5–35.
Hollows, J., 2008. *Domestic Cultures*. Maidenhead: Open University Press.
Hollows, J., and D. Bell, 2011. From 'river cottage' to 'chicken run': Hugh Fearnley-Whttingstall and the class politics of ethical consumption. *Celebrity Studies*, 2(2), pp. 178–191.
Jensen, K.B., 2013. Definitive and sensitizing conceptualizations of mediatization. *Communication Theory*, 23, pp. 203–222.

Johnson, J., and M. Goodman, 2015. Spectacular foodscapes: Food celebrities and the politics of lifestyle mediation in an age of inequality. *Food, Culture and Society*, 18(2), pp. 205–222.

Johnson, P., 2013. The geographies of heterotopia. *Geography Compass*, 7(11), pp. 790–803.

Johnston, J., and S. Baumann, 2007. Democracy versus distinction: A study of omnivorousness in gourmet food writing. *American Journal of Sociology*, 113(1), pp. 165–204.

Johnston, J., and S. Baumann, 2010. *Foodies*. New York: Routledge.

Krogager, S.G.S., K.K. Povlsen, and H.-P. Degn, 2015. Patterns of media use and reflections on media among young Danes. *Nordicom Review*, 36(2), pp. 97–112.

Krotz, F., 2007. The meta-process of 'mediatization' as a conceptual frame. *Global Media and Communication*, 3(3), pp. 256–260.

Krotz, F., 2009. 'Mediatization: A concept with which to grasp media and societal change.' In: K. Lundby ed., *Mediatization: Concept, Changes, Consequences*. New York: Peter Lang, pp. 19–38.

Krotz, F., and A. Hepp, eds., 2014. *Mediatized Worlds: Culture and Society in a Media Age*. Basingstoke: Palgrave, pp. 1–18.

Leer, J., in press. '"If you want to, you can do it!": Home cooking and masculinity makeover in *Le Chef Contre-Attaque*.' In: M. Szabo and S. Koch eds., *Food, Masculinities and Home*. New York: Bloomsbury Academic.

Leer, J., and K.M. Kjær, 2015. Strange culinary encounters: Stranger fetishism in Jamie's Italian Escape and Gordon's Great Escape. *Food, Culture and Society*, 18(2), pp. 309–328.

Livingstone, S., 2009. On the mediation of everything. *Journal of Communication*, 59(1), pp. 1–18.

Livingstone, S., and P. Lunt, 2014. 'Mediatization: An emerging paradigm for media and communication research.' In: K. Lundby ed., *Mediatization of Communication*. Handbooks of Communication Science, 21. Berlin: De Gruyter Mouton, pp. 703–723.

Lundby, K. ed., 2009. *Mediatization: Concept, Changes, Consequences*. New York: Peter Lang.

Lupton, D., 1996. *Food, the Body and the Self*. London: Sage Publications.

Maffesoli, M., 1996. *Time of the Tribes*. London: Sage.

Miller, T., 2007. *Cultural Citizenship*. Philadelphia: Temple University Press.

Naccarato, P., and K. LeBesco, 2012. *Culinary Capital*. New York, NY: Berg Publishing.

Peterson, R.A., and R.M. Kern, 1996. Changing highbrow taste: From snob to omnivore. *American Sociological Review*, 61(5), pp. 900–907.

Pink, S., and K.L. Mackley, 2013. Saturated and situated: Expanding the meaning of media in the routines of everyday life. *Media, Culture & Society*, 35(6), pp. 677–691.

Reckwitz, A., 2002. Toward a theory of social practices. *European Journal of Social Theory*, 5(2), pp. 243–263.

Reilly, J., 2006. 'The impact of the media on food choice.' In: R. Shepherd and M. Raats eds., *The Psychology of Food Choice*. Wallingford: CABI, pp. 201–223.

Rittenhofer, I., and K.K. Povlsen, 2015. Organics, trust, and credibility: A management and media research perspective. *Ecology & Society*, 20(1), p. 6. http://www.ecologyandsociety.org/vol20/iss1/art6/

Ritter, R., and B. Knaller-Vlay, 1998. *Other Spaces: The Affair of the Heterotopia*. Dokumente zur Architektur 10. Graz, Austria: Haus der Architektur.

Rojek, C., 2014. 'Big citizen' celanthropy and its discontents. *International Journal of Cultural Studies*, 17(2), pp. 127–141.

Rousseau, S., 2012a. *Media Food*. New York, NY: Berg.
Rousseau, S., 2012b. *Food and Social Media*. Plymouth: AltaMira Press.
Schatzki, T., 1996. *Social Practices: A Wittgensteinian Approach to Human Activity and the Social*. Cambridge: Cambridge University Press.
Schatzki, T., 2001. 'Practice minded orders.' In: T. Schatzki, K. Knorr-Cetina and E. von Savigny eds., *The Practice Turn in Contemporary Theory*. London: Routledge, pp. 50–63.
Schatzki, T., 2002. *The Site of the Social*. Pennsylvania: Pennsylvania State University Press.
Schulz, W., 2004. Reconstructing mediatization as an analytical concept. *European Journal of Communication*, 19(1), pp. 87–101.
Silverstone, S. ed., 2005. *Media, Technology and Everyday Life in Europa*. Aldershot: Ashgate, pp. 1–20.
Simmel, G., 1972. *Simmel on Individuality and Social Forms*. Ed. D.N. Levine. Chicago: University of Chicago Press.
Simmel, G., 2008. *Gesamtausgabe Bd. 18*. Frankfurt aM: Suhrkamp.
Thorsø, M., T. Christensen, and K.K. Povlsen, forthcoming 2016. 'Organics' are good, but we don't know exactly what the term means – Trust and knowledge in organic consumption. *Food, Culture and Society*, 19(4).
Warde, A., 1997. *Consumption, Food, and Taste*. London: Sage.

Part I

Food practices in the media

2 Good fare and welfare

Perceptions of American and French food in postwar cookbooks

Caroline Nyvang

Food is often regarded as an emblematic aspect of the "Americanization" of Europe.[1] The US has been referred to as a fast food nation and its impact on Western Europe after 1945 was dubbed a "Coca-colonization" and "McDonaldization" (Ritzer, 2004; Wagnleitner, 1994). When fares or foodstuffs are used as a synecdoche for an entire culture, it reminds us that food is a 'total social phenomenon' in the sense once formulated by Marcel Mauss (1966). Pervasive in our daily doings, the way food is obtained, prepared and consumed is central to the way we organize societies.

The US, however, was not the only cultural contender in Western Europe post-1945. As has been shown within areas such as law, philosophy and the arts, transatlantic cultural clashes were often played out during the Cold War with the US and France as proxies (Caute, 2005; Kuisel, 1993; Mathy, 2000; Whitman, 2004). This chapter argues that food, too, can be viewed as an area of cultural contestation, allowing different discussants to draw up, reflect on or reject blueprints for possible futures, and that cookbooks explicate and illustrate such visions.

In the wake of the Second World War, as overall welfare improved and basic foodstuffs took up a smaller portion of household budgets, a general interest in foreign foods emerged across Europe. In Denmark, this became evident in the increasing number of cookbooks promoting international cuisines and aiming at new target groups that had not previously been able to afford great dietary diversity.[2]

Especially from 1952, when the last wartime rationings subsided and imports increased, two distinct trajectories appeared within the cookbook genre, pointing to respectively the US and France as culinary role models. Until 1973, when the first signs of a major economic crisis appeared and Denmark formally became subject to the common food and agricultural policies of the EEC, the tussle between these two poles dominated Danish cookbooks.

By examining the ways in which authors of Danish cookbooks published in the years 1952–73 adapted recipes from France and the US, this chapter will explore how these culinary proto- and antitypes have been employed to discuss the direction of Danish food culture at the brink of a new era.[3] Besides pointing to different expectations related to gender, the chapter also shows how perceptions of American and French food served as a pretext for developing two distinct modes of consumer empowerment, underlining the broad political significance of food.

Food and the power of everyday life

During the past and present centuries, the power of everyday life has been a fulcrum to the social sciences (Highmore, 2002). The interest in everyday practices has gained ground not least in regards to twentieth century Scandinavia where the regulation of quotidian domestic habits has played a central role in the welfare policy. As documented by Swedish scholar Yvonne Hirdman, on the 'Scandinavian middle way' between socialism and capitalism, arranging seemingly mundane routines such as child rearing, love making and cleaning became key to social engineering and was ascribed immense political power (Hirdman, 1990; Hirdman and Vale, 1992).

Belonging to this realm of everyday life practices, food, too, carries a significant transformative potential. The way we organize cooking and eating has effects that reach far beyond the dinner table. According to nutritional science, what we eat helps determine whether we are healthy or sick, and how we spend our income is a key factor in balancing the national budget.

Furthermore, who shops for and prepares dinner is closely linked to division of labour within the home and society in general. In short: the way we organize our lives around dinner not only reflects but also impacts the social order at large. It is thus no wonder that food has been a fundamental ingredient in Western imagination of future societies, whether they were utopian or dystopian in character (Belasco, 2006; Madden and Finch, 2006; Stock et al., 2015).

Cookbooks elucidate this relationship between food and ideology. The dissemination of printed cookbooks across geographical as well as social boundaries has played an important role in gender, ethnic and national identity formations. By tracing how recipes have travelled between families and social layers, scholars have been able to map social networks of women in the eighteenth, nineteenth and twentieth centuries (Bower, 1997; Leonardi, 1989; Theophano, 2002). Additionally, recent research have looked into the ways recipes have been employed in the reconstruction of traditional gender roles in the wake of both world wars (Neuhaus, 2003; Novero, 2000).

In *How to Make a National Cuisine: Cookbooks in Contemporary India,* anthropologist Arjun Appadurai (1988) demonstrates how cookbooks have supported the development of a national consciousness in India. Since then, a long list of studies has followed Appadurai's lead in investigating how the dispersal of recipes has created a sense of belonging to ethnic and national communities (e.g. Gvion, 2009; Metzger, 2005; Rabinovitch, 2011).

As many of these studies suggest, the cookbook is a well-suited medium for discussing food-cultural ideologies. Thus, even if cookbooks do not necessarily reflect actual culinary practices now or in the past, they provide important clues to the societies that produced them (Fragner, 2000; Gold, 2007, pp. 11–12; Strasser, 2000, p. x).

Jeremy MacClancy once reminded us that cookbooks should not be approached as descriptive, but rather as prescriptive texts, 'giving us an inkling of the culinary worlds their authors would like to see' (MacClancy, 2004, p. 66). Following that line of research, this chapter argues that cookbooks should first and foremost be read as a testament to the visions of its creator(s). As with other food media, cookbooks allow us to get a grasp of the illusive ideals of cooking as well as the different authorities – state-sanctioned as well as private initiatives – concerned with the meaning making involved in food choices.

Postwar Danish cookbooks in a historical context

In 1945, cookbooks were far from a novelty on Danish bookshelves. During the nineteenth century, the cookbook developed its own distinct generic conventions and soon became a mouthpiece to a range of social entrepreneurs. With the addition of introductions, illustrations and indexes, cookbooks were no longer merely a repository for various recipes, but also served as a recommendation and interpretive framework for its own contents.

At the turn of the century, as the rural subsistence economy gave way to an urban monetary economy and food became a matter of public health concern, the printed cookbook became a popular medium with a wide array of new authors, including food companies, doctors and government agencies. The publication rate of cookbooks increased considerably in the course of the century, and even during periods when the book market was experiencing general slowdowns in sales cookbooks seemingly continued to be in demand across Europe (Blomqvist, 1980; Nyvang, 2013; Thoms, 1994). With individual publications matching the national sales figures for fiction blockbusters, such as *1984* (1949), *Atlas Shrugged* (1957) and *To Kill a Mocking Bird* (1960), cookbooks reached a wide potential readership.

In the late 1950s the Danish economy entered a prolonged phase of steady economic growth, allowing for the political and economic consolidation of the welfare state, and the increased economic prosperity enabled ordinary Danes to rearrange personal consumption patterns. In the booming years of 1958–73, the disposable income spent on non-essentials and services thus increased significantly, especially among the lower and middle classes, as basic needs took up a smaller portion of household budgets (Pedersen, 2010). The 1950s also saw a restructuring of the Danish retailing sector. As the last food rationings from the Second World War fell away and the import of foreign commodities increased, Danes had to accustom themselves to a wider selection of foodstuffs and large supermarkets opened up across the country, selling food in new and unfamiliar wrappings, cans and containers (Rostgaard, 2011). In Denmark, the 1950s, 60s and early 70s thus spelled radical new possibilities and challenges in terms of consumer behavior.

These changes warranted novel instructions, and cookbooks proved a promising medium. In a swift response to the new conditions, the postwar years saw a range of new books published by female domestic consultants who would travel to the US as representatives of the state or food industry. Upon returning home, these authors praised the convenience of prepackaged foodstuffs and new kitchen technologies. Contrary to this, another line of cookbooks, written mainly *by* and *for* male amateur chefs, endorsed a Mediterranean mode of cooking in which food was portrayed as a means of self-fulfillment. In the following, these two groups of authors – the female consultant and the male gourmand – serve as archetypal figures in an analysis of how Danish food-cultural ideals were adapted to meet new standards during the years 1952–73.

The domestic consultants

In the course of the two decades after the Second World War, the domestic consultant became a character paramount to the Danish cookbook genre. During the years of 1946–65, consultants authored almost one-third, a total of 79 first editions, of all cookbooks (Nyvang, 2013, pp. 116–17). Through these, female consultants would instruct and inform housewives and working mothers in matters related to nutrition, housekeeping and consumption.

The publications from the domestic consultants can be considered a second wave of the home economics movement that had swept Denmark at the turn of the century. Many of the authors of postwar cookbooks had received formal education in cookery through the privately and publicly funded home economics courses, which were held in the cities as well as the countryside from the 1890s onward, but unlike their prewar predecessors most cookbook authors publishing after the war were not engaged as teachers at these same schools. Instead, they were mainly employed by a food processing industry set on securing a stronghold in the kitchens of the Danish middle class (Nyvang, 2013, pp. 116–43).

The domestic consultants wrote their cookbooks against the backdrop of an increasing supply of convenience food. After having expanded its production lines to accommodate the increasing demand for canned goods during two almost consecutive World Wars, the food industry was looking to secure new markets in times of peace (Pollan, 2009, p. 26; Shapiro, 2004). Along with a burgeoning advertising industry, the consultants helped target the products at ordinary households. Recipes provided a subtle way of getting a foot in the kitchen door, and from 1928 state subsidization of domestic consultants' salaries allowed private companies to hire such consultants to promote different food products by means of the cookbook (Nyvang, 2013, p. 117).

The domestic consultants specifically targeted women who had a day job as well as a family to tend to. During the economic boom, Danish women's labour market participation rose significantly, and especially in the 1960s the share of married women in the workforce increased (Christoffersen, 1993, pp. 111–12). Looking to relieve implicitly female cookbook readers of some of their housework burdens, domestic consultants would propagate new dinner ideals that did not require investing great amounts of time and effort in preparations. Never really questioning the division of labour within the home, the consultants suggested that housewives and especially working mothers exercised their newfound purchasing power on timesaving convenience food such as frozen and canned food.

The American link

The enthusiasm for convenience food was unequivocally linked to ideals associated with American food culture. Cookbook author Ellen Rydelius describes her enthusiasm in this way:

> All thanks to cans and sub-zero temperatures, in half an hour the American housewife is able – if she deems it convenient – to whip up a dinner for ten people. She simply needs to mix two cans of soup, thaw a chicken, dress a lovely green salad with marinade from bottles and bake a *cake mix* – and we are beginning to learn from her.[4]
>
> (Rydelius, 1956, p. 117)

Even when the incentive to incorporate commercially prepared foods was explicitly associated with a particular American way of life, the recipes were rarely taken from US cookbooks. Most often, the American kitchen was presented to Danish readers through translations of Swedish cookbooks. The fact that US influence arrived via Sweden is consistent with previous research on the adaptations of American culture, which has shown that influential representations of America often came indirectly from within Europe (e.g. Vyff, 2011). And when it comes to cookbooks, this is hardly surprising. At a time when the pioneers Julia Child, Simone Beck and Lousette Bertholle were *Mastering the Art of French Cooking* (1961), cookbook authors within the US increasingly looked to France for culinary inspiration and all-American recipes seldom came out in print (Hoganson, 2007, p. 107). Yet it is quite telling that, out of only four American cookbooks translated into Danish during the years 1945–73, Myra Waldo's *Cooking for the Freezer* (1960) was the most successful.

Several Danish cookbook authors had direct ties to the US. In the years following the war, many of them paid visits to the US, either on research trips as part of the Marshall Plan or funded by the companies they represented. Alongside the financial support for employing consultants, the Danish state offered to partly cover travel expenses for domestic consultants, thus encouraging companies to give consultants international experience, and often the consultants would set sail for the US. For instance, the American company Tupperware invited Kirsten Hüttemeier (1914–2003) – a prominent Danish food consultant at the time and perhaps the most productive European cookbook author of the twentieth century – to the US. During her stay, Hüttemeier visited a number of food laboratories as well as test kitchens, and according to her autobiography she returned home with a fresh repertoire of recipes for quick, freezable dishes (Hüttemeier, 1977, pp. 91–92).

Likewise, Alice Bruun (1902–95), author of *Kvik mad med konserves* (1957) [Quick food is canned food], went to the US on behalf of the Danish state. As per customary procedure, she recounted her experiences in the form of a report, detailing encounters with promising oddities such as 'Fish Sticks' and 'T.V. Dinners' (Bruun, 1962). Thus, even if the Danish cookbook authors were not conveying what people in the US were actually reading and eating, the domestic consultants did indeed see certain prospects in the very epitome of all-American eating habits.

Dinnertime

By way of the cookbooks, the consultants argued that the newfound economic freedom of the late 1950s ought to be reflected in a food culture that took full advantage of the convenience offered by processed foods and new kitchen appliances. Furthermore, with an educational background in the sound cooking associated with the home economics movement, the female cookbook authors in the 1950s, 60s and 70s seemed happy to break free from their predecessors' careful attention to the cost-effectiveness of seasonal cooking. Embracing the diversity offered by tinned, frozen and vacuum-packed foodstuffs, the author of *Lidt men godt* [Little but good] enthusiastically wrote: 'Meal planning then and now differ significantly from each other, predominantly due to the fact that we – thanks to the technology – are now able to buy many fruits and vegetables all year round' (Östenius et al., 1957, introduction).[5] In equally appraising tones, the foreword to *Nem mad med konserves* [Easy cooking with preserved foods] ensured readers that canned foodstuffs even promoted a healthier diet since it had 'evened out the seasons . . . thanks to canned foods, we are now able to eat a more nutritious and diverse fare year-round' (Östenius, 1959, foreword).[6]

The fact that efficacy was a principal concern permeated the cookbooks written by domestic consultants. This was expressed in both layout and text. Clocks scattered throughout the books served as tiny iconographic reminders that time was indeed a focal point. Total preparation time was often given for each recipe and the index ordered accordingly so that the reader would be able to pick dishes by the amount of time she had on her hands. In *Nem mad for travle husmødre* [Easy cooking for busy housewives], a cookbook promising to have supper ready in as little as twenty minutes, Ruth Gunnarsen-Jensen assertively wrote: 'However busy the housewife is, dinner must be taken care of every single day, and that is why we need to find new work methods, which will reduce preparation time' (Gunnarsen-Jensen, 1957, introduction).[7]

Taking a cue from the scientific management trend of early twentieth century industrial production, rationalization was at the heart of the new work methods that Gunnarsen-Jensen and others called for. Postwar cookbook authors counted steps and suggested a reorganization of both kitchen space and family life in order to encourage time-efficient workflows (e.g. Andersen and Nielsen, 1952; Gunnarsen-Jensen, 1957). The same need for speed can also be traced in the recipe repertoire. For example, dishes such as casseroles and omelettes were novel and popular with the consultants. They were considered less laborious since the ingredients could be prepared in one batch. In the same vein, readers were also urged to save time and effort by keeping dinners to a single course.

This attention to efficiency also affected the setting of the family meal as pots, pans and even cans made their way out of the kitchen and onto the dinner table. Besides

Figure 2.1 The clock and the casserole served as recurrent iconographic reminders that convenience and time were the primary concern with the domestic consultants.

Source: Frontpage from Gunnarsen-Jensen, R., *Nem mad for travle husmødre*, 1957.

getting fewer bowls dirty, this provided a fresh décor. In *50 Gryderetter* [50 casseroles], aesthetics and rationality were brought together in the introduction: 'If you set the table with a colorful tablecloth . . . and stainless steel then ovenproof dishes and the modern casserole will fit right in', the authors promise and further assure:

> The food will keep warm, the housewife will not have to worry about decorating the dish, and many steps between the dining room and kitchen are saved as you will not need to keep 4–5 pots cooking when preparing dinner.[8]
>
> (Andersen and Nielsen, 1952, p. 4)

Exotic fruits, made available through both the influx of canned foods and the repeal of import restrictions, were also celebrated in the aesthetics of

Figure 2.2 'The consumer glasses'

Source: From Bruun, A. & K. Wonsild, *Kvik mad med konserves,* 1957.

contemporary cookbooks. Fruits were main ingredients in the numerous new salads, which accompanied the main course, and the flamboyant varieties did well in the colour photos that made their way into cookbooks during the 1950s. In particular, sliced pineapple appeared to be a favourite garnish with the domestic consultants.

In general, the visual perception of food was placed at the heart of the consultants' cookbooks. First of all, in the large self-service stores that took over from the smaller grocery shops, the housewife usually had to locate and pick the foodstuffs herself. Secondly, food was increasingly tinned, frozen or vacuum-packed, rendering the faculties of smell, taste and sensation useless for estimating the quality of foodstuffs. This paved the way for new types of instructions. With comprehensive sections on shopping and labelling, the consultants' cookbooks acted as an interface between producer and consumer.

The gourmand

During the 1960s, a number of trendsetting male artists, journalists and intellectuals began writing and publishing cookbooks. This was indeed a novel trait since – apart from the occasional medical doctor writing in the inter-war period – women had dominated the genre throughout most of the nineteenth and twentieth centuries (Nyvang, 2008).

Interestingly, male cookbook authors writing in the 1950s and 60s were predominantly self-taught and only few of them had any professional affiliation with the food industry, e.g. as restaurateurs or trained chefs. Instead, these authors were often associated with the art scene, either as performers or critics. For example, Mogens Brandt (1909–70), Olaf Ussing (1907–90) and John Price (1913–96) were renowned actors with Copenhagen theatres while, besides dishing out recipes in various media, Leif Blædel (1923–2013) was a journalist with the Danish newspaper *Information* and Jens Kruuse (1908–78) was editor of the culture section at the daily newspaper *Morgenavisen Jyllands-Posten*.

That these mediators of culture had legitimacy as food experts indicates a shift in authority, in part driven by the fact that food could now be referred to as an art form.[9] The parallel between cooking and art production was a recurrent analogy in cookbooks of the 1960s and 70s. The introduction to Mogens Brandt's *Man tager et sølvfad* [You take a silver platter] was tellingly entitled 'Ouverture' and accompanied by a drawing of the author posing as a musical conductor, waving a baton (Brandt, 1965, p. 7). Likewise, the recipes in Jens Kruuse's *En rejse værd. Gastronomisk ekspedition* [Worth travelling for. A gastronomic expedition] were defined as 'waltzes, sonatas, concertos' (Kruuse et al., 1968, p. 169).[10] And Leif Blædel – writing under the stage name of Jacques de France – even argued that the newly founded Ministry of Culture, instead of the Ministry of Agriculture, should be put in charge of Danish food policies (de France, 1964, pp. 119–20).

Figure 2.3 With the self-proclaimed gourmands as instigators, the cookbook genre took on a specific masculine expression in stark contrast to previous publications. Front-pages now featured men facing the stove and in the introductions men went shopping. For the first time in Danish cookbooks, authors thus encouraged men to put on an apron and enter the kitchen.

Source: Frontispiece from Ussing, O., *Lad os gå i køkkenet og slappe af,* 1970 [Let's relax in the kitchen]

Gastronomic ambitions

When *L'Académie de la Gastronomie au Danemark*[11] was established in 1964, the Danish gastronomic movement took shape and gained momentum. The academy was modelled on similar organizations, e.g. the Swedish *Gastronomiska Akademien* (estab. 1958) and *Gastronomische Akademie Deutschlands* (estab. 1959). Pointing out the perceived threats of quick meals and canned foodstuffs, these institutions univocally urged an intervention in the perceived downfall of European food culture (e.g. Fjellström and Haglund, 2009).

Along with features and debate pieces in the daily press, the cookbook became an important medium in these endeavours. In Denmark, "gastronomy" was a recurrent and vital concept in the introductions to cookbooks as founding members of the academy came to dominate the genre in the latter part of the 1960s and early 1970s. At the time, gastronomy was not a new concept, but nor was it particularly well-defined. Etymologically derived from the Ancient Greek words for stomach (gastēr) and regulations (nómos), the term was adopted during the neoclassicist movement of the nineteenth century to denote the norms of a distinctly French cuisine emerging in the wake of the Revolution of 1789 (for a discussion of gastronomy as a term, see Santich, 2004).

There were certainly ties and similarities between the early French and later European gastronomic ambitions. Both movements praised the epicurean experience and Danish authors of cookbooks regularly referred to forerunners from the time of the French Revolution.[12] However, in Danish cookbooks from the second half of the twentieth century, gastronomy was imbued with a particular meaning of its own. The authors viewed their cookbooks as a countermove against contemporary US inspired and commercially induced consumerism. In *Køkkenglæder* [Kitchen joy], Mogens Brandt wrote:

> I view the increasing worldwide interest in gastronomy as a protest against this plastic-and-cola culture, which is to all intents and purposes forced onto us for commercial reasons and managed by an industry as cynical as its advertising experts are clever.[13]
>
> (Brandt, 1968, p. 8)

With the gourmands, the idea of a certain culture being forced onto people was a recurrent theme, often formulated with frozen and canned foods in mind. According to *Sommermad* (1968) – a Danish translation of Elizabeth David's *Summer Cooking* (orig. 1955) – the food industries exerted a 'hypnotic power', which had consumers 'clamouring for canned pineapple', even when fresh fruit was in season (David, 1968, p. 8).[14] In general, canned fruit became emblematic in the cookbook authors' critique of the food culture promoted by the domestic consultants. Conrad-Bjerre Christensen (1914–76) morosely characterized contemporary trends in cooking as 'consultant's food with sliced tangerine, rings of pineapple and Maraschino cherries' (Bjerre-Christensen, 1967, p. 9),[15] and Mogens Brandt lamented against 'what I have dubbed "the pineapple kitchen" . . . an unsavory and foul-tasting hotchpotch, whose only true asset is that it does so well in colour-photography' (Brandt, 1968, p. 11).[16]

Often, the domestic consultants would retort, arguing that the gourmands were, in reality, appealing only to men with no children and plenty of time and money on their hands (e.g. Kaufmann, 1965; Nørgaard, 1965).

While the US with its 'consultant's food' was presented as a negative role model, the gastronomic movement heralded France as a culinary apotheosis. The recipes in the gourmands' cookbooks were often gathered from contemporary French publications or, more frequently, picked up on travels to France. It was a hallmark of the repertoire, however, that rural, provincial recipes took up most of the space. The cookbook authors, who in the 1950s and 60s looked to France for inspiration, thus separated themselves from the connotations of classic *haute cuisine,* instead pushing for a minimalistic and old-fashioned mode of cooking, which resembled the concurrent wave of nouvelle cuisine.[17]

Arbiters of taste

Refuting the idea of collective norms of taste, which epitomized the scientific ideals of the domestic consultants, the gourmands embraced the individual sense of taste. This was evident in a number of ways.

Breaking with the common conception of the gourmand as a glutton, Ejler Jørgensen (1906–83), owner of the acclaimed French restaurants *Nordland* and *Escoffier* in Copenhagen, greeted the readers of his first cookbook as: 'Dear gourmands', and went on to explain the salutation thus: 'When I say Gourmand, I am thinking of the connoisseur, he, whose palate is so developed that he knows how to differentiate, knows how to appreciate, to enjoy' (1965, p. 7).[18]

On a practical level, the development of an individual sense of taste was encouraged by recipes, which were minimalistic and less didactic compared to the conventions of the genre. For instance, several authors discarded the list of ingredients that is normally used to present the components of a dish, and the style became more anecdotal as each recipe was usually motivated by personal stories. Furthermore, the authors rarely gave specific instructions, but relied on tactile explanations. Approximate descriptions such as 'tender' and a 'dollop' thus came to replace precise cooking times and measurements.

Several authors made it clear that this was indeed a conscious style, intended to inspire the reader. Ejler Jørgensen stressed that his *Culinarisk Kalender* [Culinary calendar] did not consist of 'detailed recipes, but ideas, appealing to your fantasy and innovative powers' (1966, p. 7).[19] Tore Wretman (1916–2003) also underlined that the contents of his *Menu* should first and foremost be 'considered a guideline and an inspiration. As with any other artistic activity, cooking should not be reduced to slavish imitation' (1960, introduction).[20]

The notion of a distinctive taste also extended to the dishes, which were promoted for their use of locally grown ingredients in season. This was thought to provide each dish with a unique flavour, honouring the authentic zest of the ingredient in keeping with the premises of the nouvelle cuisine (Rao et al., 2003,

p. 817). Authors also seemed to agree that the excessive use of garnishes, sugar and even spices was an abomination as it would conceal the particular taste of the ingredients. In the words of Leif Blædel, these additives should be considered an even 'more efficient means of food destruction than the ketchup of the Britons and Americans' (de France, 1964, p. 12).[21]

In assessments such as these, the authors echoed Roland Barthes' analysis of food photography in contemporary women's magazines, which he presented in his influential text on the 'Ornamental Cookery' from 1957 (Barthes, 2000). This type of cookery, 'based on coatings and alibis . . . is for ever trying to extenuate and even disguise the primary nature of foodstuffs', Barthes wrote (2000, p. 78). In contrast to the gastronomic ambition of bringing taste to the fore, the ornamental attention to appearance produces a sense of food that, according to Barthes, belongs to 'the eye alone' (2000, p. 78).

In the cookbooks, the attenuation of the visual in favour of the gustatory had implications beyond the mode of cooking. In a world of increasing supplies and rising purchasing power, taste was suggested as the main tool for identifying the authentic. Thus, the gastronomes placed the palate as both the primary sensory organ and a critical apparatus in consumer empowerment, effectively fusing gustatory and aesthetic concepts of taste.

Perceptions of American and French food in postwar Danish cookbooks

In the years 1952–73, as the supply of culinary goods expanded and demands changed as a result of rising real incomes, consumers in Denmark were faced with a number of new opportunities regarding food choices. During this period, the Danish cookbook genre split into two distinct subgenres, each underlining different benefits of the increased standard of living. Under the pretext of different culinary ideals, authors negotiated the terms for a new national food culture and, in turn, gender roles and notions of quality.

With an unambiguous confidence in progress, the female consultant embraced the rationality of prepackaged foodstuffs and new kitchen technology. Placing efficiency as a core value in the kitchen, these authors equated cooking with work, thus reformulating domestic duties in agreement with the busy schedule of working mothers. In contrast to this, the 1960s and early 70s saw a line of cookbooks with a strong affinity for French food and marked by a considerable critique of modernity. Whereas the consultants emphasized respite from kitchen work as a step towards self-realization, the gourmands viewed the kitchen as a site for self-expression.

However, both author groups also operated within the premises of a capitalistic-liberalistic consumer society, focusing on a generic consumer in need of guidance in order to make informed and conscious choices amidst a growing and changing selection of commodities. When juxtaposing the different approaches to consumer choice presented in the cookbooks, two main strategies of empowerment appear. Putting their faith in the faculty of vision, the consultants advised their readers on how to use measurements and read labels. Rebelling against this adherence to collective guidelines, the gastronomic movement pointed to an irreconcilable conflict of interests between the industry and the consumer. In praise of private autonomy, the gourmands put the individual sense of taste at the heart of their project.

Thus, even if they do not reflect actual eating practices, cookbook authors instilled particular notions of quality, which could be extended beyond the kitchen. Food, then, is not only mediated, but can in itself be seen as a medium for the dissemination of wider political ideals.

Notes

1 For instance, Joseph Nye mentions food among the sources of soft power (Nye, 2004). For a further discussion of food as soft power, see Reynolds (2012).

2 These include cookbooks written by Danes as well translations of foreign cookbooks. However, many of the latter were published simultaneously in German, Norwegian, Dutch and Swedish and can be considered part of a shared Western European trend (Nyvang, 2013, pp. 51–54).

3 The themes and trajectories suggested in this chapter are based on a close reading of all identifiable first editions of Danish cookbooks printed in the years 1952–73, 397 in all, and is a continuation of research presented partly in a previous article exploring the semantic connotations of 'American' in Danish cookbooks (Nyvang, 2011) and partly in my Ph.D. dissertation (Nyvang, 2013). According to international bibliographies, a book must consist of two-thirds recipes in order to be considered a cookbook (Driver, 1989; Notaker, 2010). Furthermore, in compliance with the Danish National Bibliography, all publications of sixteen pages or less are considered pamphlets and have been excluded.

4 'Takket være dåser og kuldegrader kan en amerikansk husmoder i en snæver vending og, når hun ellers finder det bekvemt, lave en middag til 10 personer på ca. 1-2 time. Hun kan bare blande indholdet af to suppedåser, tø en kylling op, lave en dejlig grøn salat med marinade fra flasker og bage en *cakemix* – og vi er ved at lære hende kunsten af.' [Emphasis in original]

5 'Kostplaner før i tiden og nu til dags er væsentligt forskellige, hvilket blandt andet skyldes, at mange frugter og grøntsager – takket være tekniken [*sic*] – nu fås hele året.'

6 'Konservesmaden har udjævnet sæsonerne . . . Vi kan takket være konserves spise en mere næringsrig og forskelligartet kost året rundt.'

7 'Madlavningen skal og må klares hver eneste dag, hvor travlt husmoderen end har, derfor må vi finde på nye arbejdsmetoder, der kan nedsætte middagens tillavningstid.'

8 'Dækker man bordet med en kulørt dug . . . og rustfrit stål, passer de ildfaste fade eller den moderne stegegryde udmærket ind i billedet. . . . Maden holder sig godt varm, husmoderen behøver ikke at være nervøs for anretningen, og der spares mange trin mellem stue og køkken, da man ikke har 4–5 gryder i gang til middagen.'

9 A reminiscent form of postwar 'artistic criticism', a term originally coined by Pierre Bourdieu, is discussed in Boltanski and Chiapello (2007).

10 'Valse, sonater, koncerter'.

11 Indicative of its homage to France, the Academy has no formal English title.

12 Jean-Anthelme Brillat Savarin (1755–1826), Grimod de la Reynière (1758–1837) as well as Marie-Antoine Carême (1783–1833) are considered the seminal figures to the dissemination of the concept of gastronomy in nineteenth century France (Ferguson, 1998).

13 'Jeg ser den stadigt stigende gastronomiske interesse verden over som en protest mod denne plastic-og colakultur, som i alt væsentligt er påtvunget os af kommercielle grunde og forvaltet af en industri, der er lige så kynisk som dens reklameteknikere er smarte.'

14 'hypnotisk kraft' and 'råbe på ananas i dåse'.
15 'konsulentmad med mandarinsnitter, ananasskiver og cocktailbær'.
16 'hvad jeg har døbt "ananaskøkkenet" . . . en uappetitlig og ildesmagende pærevælling, hvis fornemste egenskab er, at den lader sig farvefotografere så smukt.'
17 The Nouvelle Cuisine, as it was presented in French restaurants in the 1960s, is analyzed in Rao et al. (2003).
18 'Kære Gourmand'er . . . Når jeg siger Gourmand, mener jeg finsmageren, den, hvis smagssans er så udviklet, at han forstår at skelne, forstår at appréciere, at nyde'.
19 'minutiøse opskrifter, men idéer, der appellerer til Deres fantasi og skaberevner'.
20 'tænkt som en vejledning og en inspiration. Madlavning bør lige så lidt som anden kunstnerisk virksomhed blive stående ved slavisk efterligning'.
21 'et endnu mere effektivt ødelæggelsesmiddel mod mad end briternes og amerikanernes ketchup'.

References

Andersen, G. and Nielsen, G., 1952. *50 gryderetter: Eenretsmiddage til hverdag og fest*. Nyt Nordisk Forlag: Copenhagen.

Appadurai, A., 1988. How to Make a National Cuisine: Cookbooks in Contemporary India. *Comparative Studies in Society and History* 30(1), pp. 3–24.

Barthes, R., 2000 (1957). *Mythologies*. Vintage: London.

Belasco, W., 2006. *Meals to Come: A History of the Future of Food*. University of California Press: Berkeley.

Bjerre-Christensen, C., 1967. *Mad jeg kan li'*. Berlingske Forlag: Copenhagen.

Blomqvist, H., 1980. *Mat och dryck i Sverige*. LTs förlag: Stockholm.

Boltanski, L. and Chiapello, E., 2007. *The New Spirit of Capitalism*. Verso: Brooklyn.

Bower, A., 1997. *Recipes for Reading: Community Cookbooks, Stories, Histories*. University of Massachusetts Press: Amherst.

Brandt, M., 1965. *Man tager et sølvfad*. Hans Reitzel: Copenhagen.

Brandt, M., 1968. *Køkkenglæder*. Hans Reitzel: Copenhagen.

Bruun, A. and Wonsild, K., 1957. *Kvik mad med konserves, og ideer til campingmad: over 200 gennemprøvede opskrifter og ideer*. Jul. Gjellerups Forlag: Copenhagen.

Bruun, A., 1962. *Rapport over Studierejse Til USA, 1960–61*. Fiskeriministeriets forsøgslaboratorium: Lyngby.

Caute, D., 2005. *The Dancer Defects: The Struggle for Cultural Supremacy During the Cold War*. Oxford University Press: Oxford.

Christoffersen, M.N., 1993. *Familiens ændring: en statistisk belysning af familieforholdene*. Socialforskningsinstituttet: Copenhagen.

David, E., 1968. *Sommermad*. Hans Reitzel: Copenhagen.

Driver, E., 1989. *A Bibliography of Cookery Books Published in Britain, 1875–1914*. Prospect Books: London.

Ferguson, P.P., 1998. A Cultural Field in the Making: Gastronomy in 19th-Century France. *American Journal of Sociology* 104(3), pp. 597–641.

Fjellström, C. and Haglund, A.-C. (Eds.), 2009. Gastronomisk Akademien och gastronomisk forskning. In: *Gastronomisk Forskning*. Gastronomiska Akademin: Stockholm, pp. 3–12.

Fragner, B., 2000. Social Reality and Culinary Fiction: The Perspective of Cookbooks from Iran and Central Asia. In: Zubaida, S. (Ed.), *A Taste of Thyme: Culinary Cultures of the Middle East*. Tauris Parke Paperbacks: London, pp. 63–71.

France, J. de, 1964. *Til bordet*. Hans Reitzel: Copenhagen.
Gold, C., 2007. *Danish Cookbooks: Domesticity and National Identity, 1616–1901*. University of Washington Press: Seattle.
Gunnarsen-Jensen, R., 1957. *Nem mad for travle husmødre: 350 opskrifter med tidsangivelser*. Jul. Gjellerups Forlag: Copenhagen.
Gvion, L., 2009. What's Cooking in America?: Cookbooks Narrate Ethnicity: 1850–1990. *Food, Culture and Society* 12(1), pp. 53–76.
Highmore, B., 2002. *Everyday Life and Cultural Theory – An introduction*. Routledge: London.
Hirdman, Y., 1990. *Att lägge livet tillrätta – studier i svensk folkhemspolitik*. Carlssons: Helsingborg.
Hirdman, Y. and Vale, M., 1992. Utopia in the Home. *International Journal of Political Economy* 22(2), pp. 5–99.
Hoganson, K.L., 2007. *Consumers' Imperium: The Global Production of American Domesticity, 1865–1920*. The University of North Carolina Press: Chapel Hill.
Hüttemeier, K., 1977. *Opskriften på mit liv*. Bulfs: Horsens.
Jørgensen, E., 1965. *Velbekomme, 100 causerier om mad*. Thaning & Appel: Copenhagen.
Jørgensen, E., 1966. *Culinarisk calender (1): Vaaren. Marts, april, maj*. Thaning & Appel: Copenhagen.
Kaufmann, L., 1965. Krydrede meninger: nye reaktioner på kronikken, der sagde "tak for mad". *Politiken*, 8 Jan, p. 12.
Kruuse, J., Aagaard, E., Aagaard, S. and Kruuse, A., 1968. *En rejse værd. Gastronomisk ekspedition, Jens Kruuse viser vej, Annabeth Kruuse giver opskrifter, Elisabeth Aagaard fortæller om vin og steder, Sven Aagaard illustrerer*. Gyldendal: Copenhagen.
Kuisel, R.F., 1993. *Seducing the French: The Dilemma of Americanization*. University of California Press: Berkeley.
Leonardi, S.J., 1989. Recipes for Reading: Summer Pasta, Lobster à la Riseholme, and Key Lime Pie. *PMLA* 104(3), pp. 340–347.
MacClancy, J., 2004. Food, Identity, Identification. In: Helen Macbeth and Jeremy MacClancy (Eds.), *Researching Food Habits: Methods and Problems*. Berghahn Books: New York, pp. 63–73.
Madden, E.M. and Finch, M.L. (Eds.), 2006. *Eating in Eden*. University of Nebraska Press: Lincoln.
Mathy, J.-P., 2000. *French Resistance: The French-American Culture Wars*. University of Minnesota Press: Minneapolis.
Mauss, M., 1966. *The Gift: Form and Functions of Exchange in Archaic Societies*. Cohen & West: London.
Metzger, J., 2005. *I köttbullslandet: konstruktionen av svenskt och utländskt på det kulinariska fältet* (PhD thesis). Acta Universitatis Stockholmiensis: Stockholm.
Neuhaus, J., 2003. *Manly Meals and Mom's Home Cooking: Cookbooks and Gender in Modern America*. The Johns Hopkins University Press: Baltimore.
Nørgaard, E., 1965. Tak for mad! *Politiken*, 4 Jan, pp. 15–16.
Notaker, H., 2010. *Printed Cookbooks in Europe, 1470–1700: A Bibliography of Early Modern Culinary Literature*. HES & De Graaf Publishers BV: Houten.
Novero, C., 2000. Stories of Food: Recipes of Modernity, Recipes of Tradition in Weimar Germany. *Journal of Popular Culture* 34(3), pp. 163–181.
Nye, J.S., 2004. *Soft Power: The Means to Success in World Politics*. Public Affairs: New York.

Nyvang, C., 2008. Mutter som kogebogsforfatter. *Personalhistorisk Tidsskrift* 1, pp. 101–113.
Nyvang, C., 2011. A l'américain – amerikansk mad i danske kogebøger 1945–70. In: Simonsen, D.G. and Vyff, I. (Eds.), *Amerika og det gode liv: materiel kultur i Skandinavien i 1950'erne og 1960'erne*. Syddansk Universitetsforlag: Odense, pp. 145–153.
Nyvang, C., 2013. *Danske trykte kogebøger 1900–70. Fire kostmologier* (PhD thesis). University of Copenhagen, Faculty of Humanities.
Östenius, A., 1959. *Nem mad med konserves: 58 opskrifter.* Thorkild Becks Forlag: Copenhagen.
Östenius, A., Olsson, B. and Borgström, L., 1957. *Lidt men godt: 30 menuer med opskrifter*. Thorkild Becks Forlag: Copenhagen.
Pedersen, J., 2010. *Danmarks økonomiske historie 1910–1960*. Multivers Academic: Copenhagen.
Pollan, M., 2009. Out of the Kitchen, onto the Couch. *New York Times*, 2 Aug, 2009.
Rabinovitch, L., 2011. A Peek into Their Kitchens: Postwar Jewish Community Cookbooks in the United States. *Food, Culture and Society* 14(1), pp. 91–116.
Rao, H., Monin, P. and Durand, R., 2003. Institutional Change in Toque Ville: Nouvelle Cuisine as an Identity Movement in French Gastronomy. *American Journal of Sociology* 108(4), pp. 795–843.
Reynolds, C.J., 2012. The Soft Power of Food: A Diplomacy of Hamburgers and Sushi? *Food Studies* 1(2), pp. 47–60.
Ritzer, G., 2004. *The Mcdonaldization of Society*, Rev. new century ed. Pine Forge Press: Thousand Oaks, CA.
Rostgaard, M., 2011. Mrs. Consumer og fremvæksten af selvbetjeningsbutikker i USA. In: Sørensen, N.A. (Ed.), *Det amerikanske forbillede? Dansk erhvervsliv og USA, ca. 1920–1970*. Syddansk Universitetsforlag: Odense, pp. 239–265.
Rydelius, E., 1956. *Fremmed mad smager bedst*. Det Danske Forlag: Copenhagen.
Santich, B., 2004. The Study of Gastronomy and Its Relevance to Hospitality Education and Training. *International Journal of Hospitality Management* 23(1), pp. 15–24.
Shapiro, L., 2004. *Something from the Oven: Reinventing Dinner in 1950's America*. Viking: New York.
Stock, P.V., Carolan, M. and Rosin, C. (Eds.), 2015. *Food Utopias: Reimagining Citizenship, Ethics and Community*. Routledge: New York.
Strasser, S., 2000. *Never Done: A History of American Housework*. H. Holt: New York.
Theophano, J., 2002. *Eat My Words: Reading Women's Lives through the Cookbooks They Wrote*. Palgrave: New York.
Thoms, U., 1994. Aufbau, Erschließung und Struktur eines Bestandes historischer Kochbücher und Haushaltslehren. *Hauswirtschaft und Wissenschaft* 42, pp. 273–280.
Vyff, I., 2011. Hvilke amerikanske drømmekøkkener? Forhandlinger af USA i dansk køkkenkultur 1950'erne og 1960'erne. In: Simonsen, D.G. and Vyff, I. (Eds.), *Amerika og det gode liv: materiel kultur i Skandinavien i 1950'erne og 1960'erne*. Syddansk Universitetsforlag: Odense, pp. 119–143.
Wagnleitner, R., 1994. *Coca-Colonization and the Cold War: The Cultural Mission of the United States in Austria After the Second World War*. University of North Carolina Press: Chapel Hill.
Whitman, J.Q., 2004. The Two Western Cultures of Privacy: Dignity Versus Liberty. *Yale Law Journal* 113, pp. 1151–1221.
Wretman, T., 1960. *Menu: en samling kulinariske optegnelser med nøje gennemprøvede opskrifter og gastronomiske anvisninger samt kommentarer.* Chr. Erichsens Forlag: Copenhagen.

3 Transcultural food and recipes for immigration

Susanna Moodie and Catharine Parr Traill

Vera Alexander

Is food made in the wilderness something to write home about? For an impoverished gentlewoman trying to settle in Canada in the 1830s it would appear to be so:

> Necessity has truly been termed the mother of invention, for I contrived to manufacture a variety of dishes almost out of nothing, while living in her school. When entirely destitute of animal food, the different variety of squirrels supplied us with pies, stews, and roasts. Our barn stood at the top of the hill near the bush, and in a trap set for such "small deer," we often caught from ten to twelve a day.
>
> (Moodie, 1962, p. 169)

There is more to this 'small fry' than meets the eye. This passage from Susanna Moodie's *Roughing It in the Bush, or Forest Life in Canada* combines an account of the settlers' living conditions with insights into a specific quotidian female creativity. Despite the hardships alluded to, the production of food is invested with imaginative potential and is shown to assume great importance for the self-image and even Ego of the narrator. While the culinary quality of the squirrel dishes is left to the gentle readers' imagination, the creative challenges posed by their preparation turn them into a small triumph. Moodie's pride in her own resourcefulness, which displaces and outshines her confession of poverty, signals a settler's survival aesthetics that is manufactured through food and disseminated through writing.

In what follows, I investigate heterotopian aspects of food writing in two prominent Canadian settlement narratives: Moodie's book from 1852, and her sister Catharine Parr Traill's *The Backwoods of Canada. Being Letters from the Wife of an Emigrant Officer, Illustrative of the Domestic Economy of British America* (1836). This chapter traces the synergies of transformation and creativity that emerge in representations of food in settlement narratives in order to show how the confection of food intersects with the invention of a new female selfhood. Moodie's English rendering of the Latin proverb 'Mater artium necessitas' offers a rare synopsis of handicraft and femininity.[1] Located in a reflection on the actual and textual production of food in transcultural learning conditions, it is symptomatic of this chapter's exploration of the connections between food, identity and textuality.

The media at the centre of this chapter are anecdotal and epistolary life narratives. Ostensibly aimed at prospective new immigrants, they document, among many other aspects of settlement, transcultural food production in colonial Canada. Catharine Parr Traill (1802–99) and Susanna Moodie (1803–85), who emigrated from Britain to Canada in 1832, found out by trial and error that life in the wilderness required skills that British gentlefolk did not generally possess. Susanna Moodie notes in her introduction: 'A large majority of the higher class were officers of the army and navy, with their families – a class perfectly unfitted by their previous habits and education for contending with the stern realities of emigrant life' (Moodie, 1962, p. xvi). Both sisters made it part of their authorial mission to educate their readers about the challenges of pioneer culture. To convey a sense of the daily challenges of living in relatively primitive conditions, the sisters fashioned narrative modes that oscillate between private and public voices and describe the mundane tasks of female domesticity in terms of risk and adventure.

While the mediation of food as a site of identity formation would appear to be a recent phenomenon symptomatic of the contingencies of globalised lifeworlds,[2] this chapter demonstrates that some of the core concerns addressed by these proceedings are already in evidence in nineteenth century settlement narratives. In many ways, the pioneer society portrayed by Moodie and Traill may be said to prefigure present-day risk societies as explored in treatises on globalisation (Beck, 1991; Appadurai, 1996; Bauman, 1998; Low, 1999; Spellman, 2002; Saussy, 2006): mobility, transformation, transcontinental communication, free market challenges, a new perception of the environment, not to forget the personal experience of a radical lifestyle change in adulthood for the emigrants make for social and cultural patterns that are comparable to later centuries – patterns for which we have only recently begun to develop frames of reference and articulation. The sisters' writings reflect the radical uncertainties of life in displacement in a manner which may be classified as heterotopian. By addressing the more remote experiences of colonial settlers, this chapter fills a gap in ongoing explorations in food cultures and transcultures. Not only does it bridge three disciplines, life writing, diaspora and gender studies, as it connects food with written constructions of place and identity, but it suggests the value of including considerations of heterotopia in food studies as a way of balancing the imaginary and material aspects of culinary writing.[3] The critical potential involved in an investigation of 'past' food and the complex implications involved in revisiting the utopian dimensions in settler narratives with a focus on food.

The acts of procuring and preparing food are connected to acts of writing. To exchange words about familiar foods and flavours is a way of strengthening bonds between people. By reporting on newly discovered food sources and unfamiliar uses of commonly known staples, and by drawing comparisons between food traditions, Moodie and Traill convey plastic, visceral impressions of their lives in a new country. Food preparation and writing are acts of sharing. Both punctuate the slow process of settlement by drawing attention to the grounding effect of repetition: food relates to daily routines. Writing about these, the immigrants wrap their extraordinary experience in habit and in a language non-immigrants can relate to: extraordinary and ordinary lifeworlds are revealed to be siblings. These daily chores of preparing food are moreover gendered and connotated as specifically female, domestic operations.

Most discussions of Moodie and Traill's works have focussed on their reinvention of notions of home, place and belonging (Thomas, 2009; Klepač, 2011) in the context of formulating a sense of Canadianness (Thurston, 1996). The status and significance of both sisters as icons of colonial femininity (Thompson, 1991; Hammill, 2003) is another dominant critical theme. Many studies have also addressed the specific genre and form of their writings, treating them variously as fiction (Freitag, 2013) or a specific form of 'colonial realism' (Whitlock, 1985). The instrumental role of food in structuring these themes is still underdeveloped and would deserve a book-length study, as would a proper investigation of the environmental concerns raised in Moodie's and Traill's writings. Investigating the tensions generated by two creative media, food production and food writing, this chapter aims to open a discussion of the complexities of creativity and femininity in the context of social and cultural transformation by addressing three intersecting thematic clusters: the heterotopian dimensions of writing food; the interactive format of life writing with its continuum of food production, mediation, and individual and collective identity formation; and transformations of femininity as articulated through food writing.

Immigrant threats and food heterotopias

While emigration is an act expressing hope for bettering one's condition, powered by textually produced images and utopian expectations, for many migrants, the actual arrival in Canada was sobering. To Susanna Moodie and Catharine Parr Traill's generation, East Coast Canada presented itself as an uncontrollable wilderness, ridden by a cholera epidemic and characterised by a rough masculine social climate. Success stories are few and far between in the life experiences of Moodie and her family in 'the bush', or 'the backwoods', as Traill wryly put it. Economic depression, forest fires, droughts, betrayal and military conflict made for an inhospitable climate for the retired British army-men and their families to try their hands at. For most, any utopian expectations of steep upward mobility through work in the colonies were quickly dashed. Such hopes became replaced by a processual dynamic of matching experiential reality and the imagination. For years, the better life envisaged by the migrants was a horizon that kept receding the more they progressed. Our heroines' husbands were military men, not farmers or traders, and hence not well-suited to the task of turning forest into farmland. Even with the best intentions and book learning, Susanna Moodie and Catharine Parr Traill could not prevent calamities caused by failed investments, underperforming crops and homestead fires. For them to write of their experiences as a model recommended for emulation was thus an ambiguous undertaking.

Colonial processes of settlement are punctuated by unsettling experiences. Not only do the settlers face the unfathomable bush; in unmooring themselves from British soil, they, too, are transformed. Social and cultural structures and hierarchies such as those of class and gender become blurry; boundaries need to be redefined. Any conventional domestic vocabulary runs the risk of euphemising the settlement experience. Both language and food production in the settlement process are located on shifting grounds.

In his influential conclusion to the 1965 *History of Canadian Literature,* Northrop Frye infamously claims that 'Canada began as an obstacle, blocking the way to the treasures of the East, to be explored only in the hope of finding a passage through it' (Frye, 1995, p. 219). In fact, so threatening is the prospect of the vast continent that Frye represents it as a monstrous leviathan: 'To enter the United States is a matter of crossing an ocean; to enter Canada is a matter of being silently swallowed by an alien continent' (Frye, 1995, p. 219). Frye's image of being engorged by the land as if it were a man-eating monster powerfully connects immigration and eating and underlines a gothic aspect food may involve. Potential threats arise from conditions of mutual change: food is transformed in the digestive process, but darkly and uncannily; food remakes and transforms the bodies of those who consume it, and potentially, their minds, thoughts, language and actions. Food representation in Traill and Moodie is framed not simply by a change of location, but by a move from the heart of civilisation to a wild periphery threatening to engulf them, transform them – one is reminded of Gilles Deleuze and Félix Guattari's point, in their essay on deterritorialisation and minor literatures, about the mouth being both the entrance point for food and the exit point for speech (Deleuze and Guattari, 1974).

From the perspective of genteel newcomers, all attempts at producing food are framed in this threat. Food must be regarded not so much as a culinary art but as part of a continuum of eating and, potentially, of being eaten. A careless wanderer in the Canadian bush might become a hungry bear's lunch. An ignorant newcomer cook might accidentally poison somebody by using an inedible herb that resembles a transatlantic relative. Food is connected to real danger. It equips the settlers for battle, but like all weapons, it may work in ambivalent ways.

Being an elementary part of the processes of acculturation, cultivation and settlement, food directs the reader's attention to enjoyable and interesting aspects of settler life and distracts from some of the hardships. Food has transformative potential in that the settlers fabricate Canada into bodily sustenance, make Canada their own by imbibing her produce. Finding and producing food and sharing the sometimes exotic discoveries about what is edible form part of a larger mission of coming to terms with an alien country and establishing control and a new stability of selfhood. Food cannot be produced once and for all: it needs to be reproduced and multiplied – it is associated with habitual, repeated acts. Step by step, day by day, meal by meal, food marks the hopes motivating the entire settlement experience. It thus provides a grid of daily routines along which human life is structured; even settler lives in which everything is otherwise 'under construction' follow this energy pattern. Beside sleep, mealtimes provide the most prominent structuring elements in our days. After breakfast, just before lunch, during dinner are landmarks on the eating map of the day, rivalling the clock. Regulatory acts of consuming food balance the conflicting demands posed by the challenge of making a home in a foreign and often hostile new environment.[4]

However, in spite of these reassuring properties, food underscores some of the uncanny aspects of immigration. The positive expectations and almost utopian hopes that motivated settlement for many of Moodie's compatriots are challenged by destabilising experiences: to set foot onto new land is menacing; how much more so to allow said land to enter and transform one's own body. The internalisation of Canadian plants, Canadian animals and Canadian ways of processing food, many of them unknown, are deeply threatening prospects. Even regardless of migration, food consumption is inherently mystical: invisible, happening in darkness inside the body, the transformation of nutrients is part of nature, an uncontrollable force. Reflection on food leads to insights about the transience of time: food exerts its power in consumption, in vanishing. Being a form of ephemeral transformative materiality, food is realised as it vanishes, it gives life as it dies. One need not spell out the overtones of Christian transubstantiation to see that food is literally a matter of life and death. Acts of writing balance and contain these threats as any textual records have the potential to outlast their writers. By writing about the processing of food, crucially, by putting names to plants and giving a date and setting to the acts of appropriating the land, Moodie and Traill establish a fragile sense of order and control.

Written food in Traill and Moodie can be seen as an expression of heterotopian ambivalence: in its promise of enjoyment and deferred delight and in its semi-fictional or textual and semi-real status, food bears some of the characteristics which Foucault attributes to spaces he labels as heterotopias: real, quasi-utopian liminal sites which mirror and subvert other spaces (Foucault, 1986, p. 24). Like the mirror which Foucault uses to illustrate his idea of heterotopia, food is involved in projections and reflections, being the material from which the settlers' bodies receive sustenance. Like Foucault's mirror, it transposes the human body and elicits uncertainty as to its exact location and composition by destabilising trust in the integrity and solidity of human life-frames. Although food is not a space as such, it can be understood as an abstract correlate of the symbolic 'sites' assembled by Foucault. In its complex affiliation with space and displacement (derived from the land, food is displaced in preparation and transforms as it 'travels' through the body) it may even be conceptualised as a contact zone of sorts, adapting Marie Louise Pratt's term for the 'interactive, improvisational dimensions of imperial encounters' to accommodate material and environmental aspects (Pratt, 2008a, p. 8): in making food, materials and competencies from different cultural and spatial origins combine. In consuming food, people of different backgrounds come together. In representing food, writers and readers collaborate to bridge the realms of the imagination and experiential reality. The evanescence of food is counteracted by the enduring creative corollary of food writing: more than one hundred years after the demise of the squirrels depicted above, readers can still absorb its textual vestiges. By vicariously reliving the trials and tribulations of Moodie and

Traill, they are induced to participate in an interactive dialogue which leads them into a diverting maze of real and confected lifeworlds.

Food and life writing

The combination of utopian imaginary dimensions and heterotopian, unsettling aspects makes the subject of written food representation into an intriguing case for a negotiation of identity formation.

Moodie and Traill are exponents of a literary family. From their teens onwards, all but one of the siblings helped to supplant the family's scarce income after their father's death in 1818 by publishing children's books, didactic stories, poems and historical biographies. Traill and Moodie wrote several books based on their experiences[5] and have since become icons of pioneer life. They have in the words of Ann Boutelle:

> found a second literary immortality in the 1970s: Moodie in Atwood's *The Journals of Susanna Moodie* (a poetic sequence); Traill in Laurence's *The Diviners* (a novel). A full century after the creation of their individual (and contradictory) myths, each sister succeeded in becoming one.
>
> (Boutelle, 1986, p. 13)

Despite this shared background and parallel fates, the sisters produced quite different accounts of settling in Canada, as prompted by their individual interests and attitudes. While Catharine is known for her knowledge of nature and botany, as evidence by several plant studies, Susanna was involved in humanitarian work for the Anti-Slavery society for which she published several tracts.[6] Susanna Moodie's memoir came out some fifteen to twenty years after the events depicted and thus offers heavily processed experience. Her practical emigration tips are embedded in colourful character sketches and witty dialogues. She complains quite engagingly about her hard luck and enlivens her narrative with little poems. By contrast, Catharine, who had even begun to write about Canadian emigrant life before actually leaving England, generates a greater sense of immediacy by publishing her edited letters to her family and arranging them to form an emigrant's diary. Catharine's description of the skills required of pioneering women encompasses various dimensions of food production:

> The female of the middling or better class [. . .] pines for the society of the circle of friends she has quitted, probably for ever. She sighs for those little domestic comforts, that display of the refinements and elegancies of life, that she had been accustomed to see around her. She has little time now for those pursuits that were ever her business as well as amusement.
>
> The accomplishments she has now to acquire are of a different order: she must become skilled in the arts of sugar-boiling, candle and soap making, the making and baking of huge loaves, cooked in the bake-kettle [. . .]. She must know how to manufacture hop-rising or salt-rising for leavening her bread;

> salting meat and fish, knitting stockings and mittens and comforters, spinning yarn in the big wheel [. . .].
>
> The management of poultry and the dairy must not be omitted; for in this country most persons adopt the Irish and Scotch method, that of churning the *milk*, a practice that in our part of England was not known. For my own part I am inclined to prefer the butter churned from cream, as being most economical.
>
> (Traill, 1836, p. 151)

Notably, one of the first skills in Traill's list is to make (maple) sugar, which takes primacy over 'candle and soap making'. This opening on a sweet luxury article gently eases readers into a gradual realisation of just how menial the female pioneer's tasks will be.

Between them, Moodie and Traill create a vivid tapestry of ways of inducing readers to learn from the mistakes they made. To differing degrees, the sisters' writings confront readers with something Paul John Eakin terms a 'referential aesthetic':[7] they depict experiences mutually understood to be real, lived, based on history and facts. Representing personal experience, both women's narratives fall under the heading of life writing. This genre is marked by a degree of plurality and instability: 'life writing is not a fixed term, and [. . .] it is in flux as it moves from considerations of genre to considerations of critical practice' (Kadar, 1992a, p. 3). Claire Lynch underlines the plural character of the term when she describes it as an attempt to 'control' a broad and 'unruly genre' which 'includes autobiographies, biographies, case studies, diaries, memoirs, autobiographical novels, ethnography, blogs, profiles and numerous other forms' (Lynch, 2009, p. 210).[8]

Like autobiography, life writings are self-conscious about their building materials and principles of construction, but in contrast to the term 'autobiography', 'life writing' privileges marginalised voices, giving a space to the quotidian and revaluing social and historical contributions of those not traditionally in power, at least in the usage feminism has popularised (Stanford Friedman, 1988; Conway, 1998; Cosslett et al., 2000). Life writing may challenge hierarchies and authorities, as David Amigoni argues: ' "Life Writing" focuses on practises of inscription, reading and interpretation that invite us to suspend and re-examine the historically received functions of, and hierarchical relations between, genres' (Amigoni, 2006, p. 3). The recent critical tendency to prioritise life writing over autobiography or biographical writing attests to the ability of this collective denominator to accommodate a plural, flexible sense of selfhood rather than any monolithic, traditionally white male dominated writing subject, as Shari Benstock argues, distinguishing female autobiographic voices from the male canon:

> For Gusdorf, autobiography 'is the mirror in which the individual reflects his own image' (33). [. . .] This definition of autobiography overlooks what might be the most interesting aspect of the autobiographical: the measure to which 'self' and 'self-image' might not coincide, can never coincide in language – not

> because certain forms of self-writing are not self-conscious enough but because they have no investment in creating a cohesive self over time.
>
> (Benstock, 1988, p. 15)

From the 1990s onwards, autobiography criticism's early focus on representing independent and delimited selfhoods, here short referenced by the allusion to Gusdorf (1980), is increasingly being replaced by an investigation of the relational interstices. Paul John Eakin replaces the subject represented in 'literature of the first person' by the phrase 'relational selves' (Eakin, 1999, p. 43) as a way to challenge autobiography's traditional focus on autonomous selfhood. Conceiving 'of "self" less as an entity and more as a kind of awareness in process' (Eakin, 1999, p. x), he suggests patterns which account for the role of relationships in forming and transforming identities. He adopts a focus on relationships and significant, 'proximate Others' in the lives of autobiographical writers from feminist autobiography criticism. In Eakin's reading, identity is necessarily pluralistic and dynamic, and it is variously shaped by others, both experientially and in the written representation on the page. It is through others that the self is realised, comes into being and is constructed as individual.

Though subject to the limitations of representation in that the writers, for reasons of space and propriety, curtail their accounts and make personal choices about what to omit and how to adorn certain memories, their works refer to actual places and events. The trustworthiness of their accounts is predicated on the self-images they project on and between the lines. These self-images include a reliable sense of respectability, class consciousness and gender propriety. Personalisation generates reliability even if what is shared is a secret, and reliability is high on the writers' agendas, chiefly in view of the abundance of propaganda material about Canadian emigration they write against. Such promotional texts are referred to in Susanna Moodie's introduction, which characterises the emigration fashions of her day and age as being subject to a misleading rhetoric and false promises:

> From the year 1826 to 1829, Australia and the Swan River were all the rage. No other portions of the habitable globe were deemed worthy of notice. These were the *El Dorados* and lands of Goshen to which all respectable emigrants eagerly flocked. Disappointment, as a matter of course, followed their high-raised expectations. Many of the most sanguine of these adventurers returned to their native shores in a worse condition than when they left them.
>
> In 1830, the great tide of emigration flowed westward. Canada became the great land-mark for the rich in hope and poor in purse. Public newspapers and private letters teemed with the unheard-of advantages to be derived from a settlement in this highly-favoured region.
>
> (Moodie, 1962, p. xvi)

Moodie's warning metatextually introduces the matter of truth versus fabrication and makes this into a programmatic issue. The poem with which Moodie dedicates her book to her sister Agnes Strickland mentions food as organically connected to the land and the settlement process, as the following extract shows:

> Joy, to stout hearts and willing hands,
> That win a right to these broad lands,
> And reap the fruit of honest toil,
> Lords of the rich, abundant soil.
>
> Joy, to the sons of want, who groan
> In lands that cannot feed their own;
> And seek, in stern, determined mood,
> Homes in the land of lake and wood,
> And leave their hearts' young hopes behind,
> Friends in this distant world to find; . . .

Food is a metonymy of life and prosperity. However, the notion of living in a new earthly paradise or promised land conjured up in this rhyme holds ambiguities. Food is as much a necessity as it is a stimulant. Underlying and powering all representation is its significance as a building block of human life and its role as the foundation of community: sharing is key to both food consumption and representation, the writing about food and how to make it. But community is constituted by rules and conventions. For the upper middle class women who emigrated to Canada in search of a better life, food became one of the central ways of upholding standards at least in one corner of life.

Moodie and Traill's combination of private and public dimensions makes space for a relationality that encompasses the reader. Passages dealing with the making and consumption of food highlight the relational qualities of both food and textuality. Food is an instrumental part of evoking a sense of mutual belonging as the family members send letters enclosing seeds along with memories, recipes and ideas. In the ninth of the eighteen open letters which make up *The Backwoods of Canada,* Catharine Parr Traill addresses to her relatives a number of requests:

> The first time you send a parcel or box, do not forget to enclose flower-seeds, and the stones of plums, damsons, bullace, pips of the best kinds of apples, in the orchard and garden, as apples may be raised here from seed, which will bear very good fruit without being grafted; the latter, however, are finer in size and flavour. I should be grateful for a few nuts from our beautiful old stock-nut trees. Dear old trees! how many gambols have we had in their branches when I was as light of spirit and as free from care as the squirrels that perched among the topmost boughs above us. "Well," you will say, "the less that sage matrons talk of such wild tricks as climbing nut-trees, the better." Fortunately, young ladies are in no temptation here, seeing that nothing but a squirrel or a bear could climb our lofty forest-trees. Even a sailor must give it up in despair.
>
> I am very desirous of having the seeds of our wild primrose and sweet violet preserved for me; I long to introduce them in our meadows and gardens. Pray let the cottage-children collect some.

> My husband requests a small quantity of Lucerne-seed, which he seems inclined to think may be cultivated to advantage.
>
> (Traill, 1836, pp. 124–125)

Catharine's food-related 'bucket list' intertwines plans for the future with personal reminiscences about a shared past. By underscoring the growth of the plants described, she alludes to her own history of growth, thereby creating a sense of continuum between the two countries and investing prospective colonial food with emotional value. The learning process she requires of her readers is connected to her own personal history of transcultural learning.

There are numerous reasons why Traill's and Moodie's writings appeal to a present-day audience. Not least of these is the instability of the texts and multiplicity of narrative modes they evince. Susanna Moodie's books went through a large number of editions, leaving today's readers with a degree of uncertainty about any authoritative version (Klinck, 1962, p. xii). Whereas Susanna Moodie enjoys a greater recognition for her literary talents, pragmatic Catharine may be said to have been a greater success as an immigrant guide. Traill gives her readers a no-nonsense account of settlement, exhorting them to not expect many comforts for the first five or so years into their Canadian existence. But ultimately both offer their writings as self-help books containing recipes for emigration. They thereby invite readers to use the narratives as scripts for their own lives and to complete them by realising the textual outlines on the stage of life.

Readers are not simply subjected to didactic lists of Dos and Don'ts but spurred into correspondence. In her eleventh letter, for instance, Traill stresses interactive traits when she frames her essay on different crops and methods of preparation as a response to concrete requests for information from her friends at home: 'You wish to know something of the culture of Indian corn, and if it be a useful and profitable crop' (Traill, 1836, p. 153), she writes, sketching an intriguingly elusive transatlantic dialogue on which readers get to eavesdrop. In their openness to responses and invitations of feedback, both sisters stress the workshop character of their writing as a sign of the processual nature of emigration as a collaborative work in progress.

Lest the relative failure of their own emigration and the numerous hardships recounted make for depressing reading, both sisters have devised ways of softening the negativity and didacticism of their narratives. Given Traill's predilection for botanical matters, which is especially pronounced in her fourteenth letter, it is unsurprising that her description of home-grown food borders on the poetic:

> Our rice-beds are far from being unworthy of admiration; seen from a distance they look like low green islands on the lakes: on passing through one of these rice-beds when the rice is in flower, it has a beautiful appearance with its broad grassy leaves and light weaving spikes, garnished with pale yellow green blossoms, delicately shaded with reddish purple, from beneath which fall three elegant straw-coloured anthers, which move with every breath of air or slightest motion of the waters.
>
> (Traill, 1836, p. 193)

Traill takes her readers on a progressive route through her field as emphasised by verbs suggesting movement both of the eye of the beholder and the colours and forms evoked. Her rice is so much more than a staple: what grows between the lines of Traill's description is a dynamic paradisal landscape that conjures up a beautiful image of the youth and prospective fruitfulness of the land she has planted.

Not that this word-painting is the only instance of pictorial art: whereas Moodie decorates her emigration account with poems, Catharine enlivens her writing with botanical drawings, maps and sketches of landscapes, animals and indigenous babies. These artworks heighten the book's aesthetic appeal and offer readers of all ages a complementary second way of imaginatively inserting themselves in the plots outlined. In part Traill's drawings emphasise processuality in representing emigration as an open-ended process. But they crucially serve the more indirect purpose of asserting the writer-artist's gentility: sketching was an acceptable feminine occupation that would reassure readers back home that the writer in question had not lost her finer sensibilities and could still be looked to as a reliable role model.

Translations of female selfhood

The ambivalence of voice and the unstable construction of identity that emerges in writings like Moodie's and Traill's make them into popular cases for an examination of modern female selfhood. Canada is their workshop for the invention and consolidation of a new female settler identity. It is in passages dealing with the preparation of food that their engagement with their new land and its possibilities comes across as an invitation to women readers to shoulder their share of the coloniser's burden and contribute to realising the utopian hopes projected onto the Canadian clearings.

The immigrant lady writers face a delicate diplomatic mission: they must show recognition of the values and accomplishments of their prospective readers while dishing out the pedagogical message that they will have to jettison or substantially transform most of them. Too radical a challenge of class conventions might put the sisters at risk of losing status. To both, emigration offered a significant increase in mobility and personal freedom. In Moodie's above-cited account of squirrel dishes, it is unclear whether the task of hunting for small animals falls to women or men, but in subsequent passages Susanna gleefully recounts going fishing with her husband, indicating the fall of conventional gender divides. Worst of all, she clearly relishes their shared outdoors activities, and, dare we suggest, her enhanced opportunities for wielding clout. Driven by her husband's attested unfitness as a farm manager (Thomas, 2009, p. 110), she extends her competencies into previously male domains. Had her contemporaries had the term 'heterotopian' at their disposal, this breach of order might well have elicited its use.

Similarly, under the motto 'Forewarned, forearmed', Catharine Parr Traill carefully outlines her motivation 'to afford every possible information to the wives and daughters of emigrants of the higher class who contemplate seeking a home amid our Canadian wilds'. Her introduction emphasises the importance of women as domestic managers:

> Among the numerous works on Canada that have been published within the last ten years, with emigration for their leading theme, there are few, if any, that give information regarding the domestic economy of a settler's life, sufficiently minute to prove a faithful guide to the person on whose responsibility the whole comfort of a family depends – the mistress, whose department it is "to haud the house in order." Dr Dunlop, it is true, has published a witty and spirited pamphlet, "The Backwoodsman," but it does not enter into the routine of feminine duties and employment, in a state of emigration. Indeed, a woman's pen alone can describe half that is requisite to be told of the internal management of a domicile in the backwoods, in order to enable the outcoming female emigrant to form a proper judgment of the trials and arduous duties she has to encounter.
>
> (Traill, 1836, p. 9)

In a postcolonial context, the word-field of ownership, possession, entitlement and belonging opens up a contradictory cluster of related, overlapping and contradictory concepts. Texts from former settler colonies articulate what Graham Huggan and Helen Tiffin call a

> crisis of belonging which accompanies split cultural allegiance, the historical awareness of expropriated territory, and the suppressed knowledge that the legal fiction of entitlement does not necessarily bring with it the emotional attachment that turns "house and land" into home.
>
> (Huggan and Tiffin, 2010, p. 82)

For the actual 'home-makers', this crisis has another dimension.

Displaced notions of female grace are among the chief challenges the immigrant ladies from England struggle with. Genteel femininity is useless in a rough social climate dominated by lumberjacks, as Moodie suggests by dismissing Canada as 'the bush'. As always, her sister is more moderate in her portrayal and reminds readers that she forms part of a second wave of immigrants who travel through the early coastal settlements on their way to the interior and benefit from the experiences of earlier arrivals. Although Traill and Moodie left England twenty years before the publication of Coventry Patmore's iconic poem about the 'angel in the house' as an ideal of Victorian womanhood, the notion of women as housebound and dependent on male guidance already existed (Perkin, 1995; Gorham, 2012). However, the demands of domesticity change radically if the actual buildings these notions are tied to still need to be built. In pioneering conditions the entire concept of the home, the house and *oikos* needed to be rethought. As Christa Zeller Thomas points out (with reference to Moodie), both house and home are unstable signifiers (Thomas, 2009).

In their new surroundings, the Victorian ladies are faced with the challenge of reinventing their femininity as something virtually unheard of: they need to become practical. The kitchen, or what passes for such in the comparatively primitive living conditions they are transported to, symbolises this shift. It is the location of their work and the place where they confront the conflicting implications of becoming workers rather than supervisors of working cooks and maids, an experience which proves quite invigorating at times, especially in the absence of genteel disapproval from their neighbours. Catharine and Susanna variously employ humour or insert conventional signposts of feminine productivity, such as poems and drawings, to forestall any misinterpretation of their continuing propriety and their authority as civilised and civiliser.

The immigrants' letters home bridge an experiential gap on a par with the geographical one of crossing the Atlantic. It is here the letters unfold their transcultural potential in the sense of Marie Louise Pratt's contact zones (Pratt, 2008b, pp. 1–12). Traill and Moodie find themselves in the position of having to justify why they take such an active interest in food production. Catharine reveals herself to be quite anxious about notions of class early into her stay: in her entry dated Sept. 11, 1832, she explains the social rank of a storekeeper as the Canadian equivalent of a member of Parliament:

> We have experienced some attention and hospitality from several of the residents of Peterborough. There is a very genteel society, chiefly composed of officers and their families, besides the professional men and storekeepers. Many of the latter are persons of respectable family and good education. Though a store is, in fact, nothing better than what we should call in the country towns at home a "*general shop*," yet the storekeeper in Canada holds a very different rank from the shopkeeper of the English village. The storekeepers are the merchants and bankers of the places in which they reside. Almost all money matters are transacted by them, and they are often men of landed property and consequence, not infrequently filling the situations of magistrates, commissioners, and even members of the provincial parliament.
>
> (Traill, 1836, p. 73)

This explication is as much for the benefit of her family as for herself: traditional categorisations of society no longer apply and there is unsung gentility in humble guises for both women and men. Many of Traill's acts of cultural translation require such a rethinking of social categories, class and status. Mobility requires a certain flexibility of mind, and in correcting conservative notions about propriety, Traill must not put her own respectability in any doubt. The relaxation of class structures occurring in the colony poses special threats for women. In a time and social climate where woman travellers tended to be frowned upon as being potentially 'loose', migrant women had to go to some length to emphasise their continuing respectability (Smith, 2001; Campbell, 2002).

Despite such constraints, a certain measure of satiric exposure comes to the fore when Traill lampoons some of the fixed ideas about class, rank and propriety she encounters in her female fellow immigrants unable to take themselves out of England:

> I was once much amused with hearing the remarks made by a very fine lady, the reluctant sharer of her husband's emigration, on seeing the son of a naval officer of some rank in the service busily employed in making an axe-handle out of a piece of rock-elm.
>
> "I wonder that you allow George to degrade himself so," she said, addressing his father.
>
> The captain looked up with surprise. "Degrade himself! In what manner, madam? My boy neither swears, drinks whiskey, steals, nor tells lies."
>
> "But you allow him to perform tasks of the most menial kind. What is he now better than a hedge carpenter; and I suppose you allow him to chop, too?"
>
> "Most assuredly I do. That pile of logs in the cart there was all cut by him after he had left study yesterday," was the reply.
>
> "I would see my boys dead before they should use an axe like common labourers."
>
> (Traill, 1836, pp. 116–117)

Being a cognate of 'domestic', 'menial' stresses the crossing of a class divide that seems an equivalent of crossing the Atlantic for some immigrants and indicates the magnitude of the didactic task at hand.

The 'most menial tasks' for women in this transcultural shift revolve around food. From the perspective of the two women writers, domestic chores relating to food production become a central location where the necessary paradigm shift from leisure class member to useful immigrant is negotiated:

> Since I came to this country, I have seen the accomplished daughters and wives of men holding no inconsiderable rank as officers, both naval and military, milking their own cows, making their own butter, and performing tasks of household work that few of our farmers' wives would now condescend to take part in. Instead of despising these useful arts, an emigrant's family rather pride themselves on their skill in these matters. The less silly pride and the more practical knowledge the female emigrant brings out with her, so much greater is the chance for domestic happiness and prosperity.
>
> (Traill, 1836, p. 150)

Food production is designated as a female domestic occupation; however, food consumption is where the sexes meet and where the male immigrants are dependent on the women's expertise, ingenuity, communicative skills, economy and imagination. The food producing activities are the covert foundations of the strength which the more strenuous and more physically demanding tasks of clearing the forest and constructing settlements rest on. Whereas Victorian society in England encoded these interconnections in social forms and class, the immediacy of life in Canada exposes them. This is an extremely empowering discovery for the settler women to make. Even though Moodie and Traill face many trials and fail to achieve the financial stability they aimed for, their work with food, botany, child raising and other domestic occupations as well as writing provide them with a notion of womanhood that highlights strength over weakness.

It is a combination of actual and textual food that constitutes the women writers as settlers. While the actual food sustains their bodies, the textual food sustains their self-images. Food and letters are interlocking parts of a practical and complex process of identity formation. Therefore it would not be far-fetched to read their writings as autobiographical variants of a female *bildungsroman,* a genre which has significant overlaps with life writing in that it recounts the personal growth and tribulations of an individual.[9] This traditionally male genre is a fitting template not least because, as discussed above, settlement demands the ladies venture into domains and activities traditionally connotated as masculine, both in actual locations and on the abstract plane of textuality. In their writings, the sisters exert a potentially subversive influence on how conventional gender boundaries are defined. Traill in particular relishes the sovereignty over botanical matters emigration affords her.[10] Kitchen and food concerns are convenient means of masking her unfeminine enthusiasm for the adventure of scientific discoveries.[11] Carefully framing her botanical exploits in a food-related civilising mission, Catharine barely bothers to restrain her glee that all male botanists, to whom she ascribes conservative ideas of social propriety, are safely contained in the old world:

> I am anxiously looking forward to the spring, that I may get a garden laid out in front of the house; as I mean to cultivate some of the native fruits and flowers, which, I am sure, will improve greatly by culture. The strawberries that grow wild in our pastures, woods, and clearings, are several varieties, and bear abundantly. They make excellent preserves, and I mean to introduce beds of them into my garden. There is a pretty little wooded islet on our lake, that is called Strawberry island, another Raspberry island; they abound in a variety of fruits, wild grapes, raspberries, strawberries, black and red currants, a wild gooseberry, and a beautiful little trailing plant that bears white flowers like the raspberry, and a darkish purple fruit consisting of a few grains of a pleasant brisk acid, somewhat like in flavour to our dewberry, only not quite so sweet. The leaves of this plant are of a bright light green, in shape like the raspberry, to which it bears in some respects so great a resemblance (though it is not shrubby or thorny) that I have called it the "trailing raspberry."
>
> I suppose our scientific botanists in Britain would consider me very impertinent in bestowing names on the flowers and plants I meet with in these wild woods: I can only say, I am glad to discover the Canadian or even the Indian names if I can, and where they fail I consider myself free to become their floral godmother, and give them names of my own choosing.
>
> (Traill, 1836, p. 120)

The project of garden-making in this passage signals the controlling power of civilisation: wilderness is being refined through feminine craft.[12] A useful rather than ornamental garden flags the practical considerations of a settler woman using her ingenuity to provide sustenance. All the demands of propriety thus satisfied, her appropriation of a male stance is well couched in safety measures. Traill's painterly joy in detail and the womanly image projected onto the fruit in question, a raspberry that trails the ground like a Victorian lady's skirt, take the brunt of the breach of social order she perpetrates in assuming an Adamic imperialist's authority over names.

Food is of course a means of survival and necessarily simple, given that it is produced on land only just claimed from the forest and prepared with comparatively primitive methods and tools, but it is also the projection ground of hopes and aspirations, as in the following passage from Traill's letters:

> I have seen some good specimens of native cheese, that I thought very respectable, considering that the grass is by no means equal to our British pastures. I purpose trying my skill next summer: who knows but that I may inspire some Canadian bard to celebrate the produce of my dairy as Bloomfield did the Suffolk cheese [. . .]. You remember the passage, – for Bloomfield is your countryman as well as mine, – it begins:
>
> "Unrivalled stands thy county cheese, O Giles," &c.
>
> (Traill, 1836, p. 153)

'Respectable' is an odd epithet for a cheese – Traill's anthropomorphic animation conflates the respectability of the cheese maker with her produce. The details of bad grass and unhappy cows serve to highlight a sporting challenge to cheese manufacture: whoever achieves a decent product under such conditions must be quite some cheese-maker. Her cheese writing is clearly an inspiration for future ambitions in the emerging field of dairy literature. Quick to be consumed, the cheese has gained comic immortality through writing, and textuality, even imagined textuality, creates an imaginary correlation between maker and her cheese. Traill's cheese has a utopian dimension of hope and yet it can be said to carry an unsettling heterotopian aspect due to the discrepancy between the idea and the material product: that which is needed to survive is not yet present; it is displaced onto textual planes.

But beside such philosophical musings on potential food, the sisters share concrete material manifestations of female gentility in food and drink. From this rare instance where Traill's humour outstrips her sister's, we turn to a (similarly rare) case where Moodie purloins Traill's role as botanical connoisseur. Here is how she depicts her family's precarious situation three years into their Canadian sojourn:

> I will give a brief sketch of our lives during the years 1836 and 1837.
>
> [. . .] All superfluities in the way of groceries were now given up, and we were compelled to rest satisfied upon the produce of the farm. [. . .] As to tea and sugar, they were luxuries we could not think of, although I missed the tea very much; we rang the changes upon peppermint and sage, taking the one

> herb at our breakfast, the other at our tea, until I found an excellent substitute for both in the root of the dandelion.
>
> The first year we came to this country, I met with an account of dandelion coffee, published in the *New York Albion*, given by a Dr Harrison, of Edinburgh, who earnestly recommended it as an article of general use.
>
> [. . .] During the fall of '35, I was assisting my husband in taking up a crop of potatoes in the field, and observing a vast number of fine dandelion roots among the potatoes, it brought the dandelion coffee back to my memory, and I determined to try some for our supper. Without saying anything to my husband, I threw aside some of the roots, and when we left work, collecting a sufficient quantity for the experiment, I carefully washed the roots quite clean, without depriving them of the fine brown skin which covers them, and which contains the aromatic flavour, which so nearly resembles coffee that it is difficult to distinguish it from it while roasting.
>
> I cut my roots into small pieces, the size of a kidney-bean, and roasted them on an iron baking-pan in the stove-oven, until they were as brown and crisp as coffee. I then ground and transferred a small cupful of the powder to the coffee-pot, pouring upon it scalding water, and boiling it for a few minutes briskly over the fire. The result was beyond my expectations. The coffee proved excellent – far superior to the common coffee we procured at the stores.
>
> [. . .] For years we used no other article; and my Indian friends who frequented the house gladly adopted the root, and made me show them the whole process of manufacturing it into coffee.
>
> [. . .] Few of our colonists are acquainted with the many uses to which this neglected but most valuable plant may be applied.
>
> (Moodie, 1962, pp. 168–169)

While she clearly has a great deal to complain about, Moodie does not dwell on the negative. Her account shifts from scarcity and debt to the joys of innovation. Deprivation leads her to find something which has relevance to readers on both sides of the Atlantic. Moodie's dandelion treatise extends over several more paragraphs, in which she details further properties of the 'super food' she has discovered, including, as an extra bonus, a recipe for dandelion beer. In contrast to the squirrel stew passage cited earlier, this one takes time to actually praise the flavour of the invention described, and by extension, the ingenuity of its inventress: poor she may be, but thanks to her genteel background and habits of self-improvement, she is acquainted with luxury stimulants such as tea and coffee, which distinguishes her from indigenous society, and thanks to her education and reading, she has something to teach her less-educated fellow-emigrants. Thus the coffee-replacement with native plants rather than exotics defines her as capable of surviving with her gentility intact, as somebody who refines and improves the diet of her fellow Canadians, managing the double role of pioneer and gentlewoman.

The passage reflects colonial notions of power and knowledge: while the Moodies' economic status has not improved significantly three years into their stay in the promised land, Susanna still defines herself as superior coloniser: she has the authority to write about the land. Her book learning outstrips the competencies of native Canadians. Despite being a newcomer, she has knowledge to impart to her 'Indian friends'; her food reading and food production position her as an informed mediator at the centre of old and new Canadians, settlers and First Nations.

A close reading of this passage reveals more than the simple assurance that settlement is potentially enjoyable. The implications for women are particularly suggestive. As a roadside flower, the dandelion is not particularly reputable. Both sides of the Atlantic, it is common as a weed, as evinced in the pejorative nicknames bestowed on it in several European languages, many referring to dogs or making unappetising connections between the flower's yellow colour and urine. Nonetheless, in Moodie's story, it becomes an evocative symbol of femininity, being invested with both signs of marginalisation and strength. The latter is implied by its poetic name, a corruption of the French *dent de lion*. It is shown to have extraordinary powers untapped in British everyday life. While this discovery takes place in the wilderness of Canada, reassuringly, it is sanctioned by the scholarship of a Scottish doctor.

The example of the dandelion coffee shows the close interplay of textuality, the imagination and concrete material reality: Susanna Moodie first comes across the idea of grinding dandelion roots in a text and she both follows the script by making the coffee and republishes the recipe to her readers, enriched with personal memories. The similarity between her act of sharing, in words if not verbal images, her culinary enterprises can be seen as an early manifestation of the present-day craze to document restaurant meals via digital photographs, a partial act of sharing which transcends the ephemeral nature of culinary art, whether the food in question is a curiosity or a prize-winner.

In contrast to her sister, Susanna Moodie's botanical expertise is more overtly in line with Victorian femininity: flowers are a female domain, and her using the dandelion for kitchen purposes is quite acceptable and ladylike. And yet, Susanna writes about her first attempt at coffee-production in terms of an experiment, stressing inventiveness, risk-taking and scientific progress. The fact that she gets the coffee idea from an old world medical man aligns food and medicine. As for readers in the old world, they are treated to an exotic tale of an adventurous nature. Dandelion coffee helps create her public image if not identity. It also offers an interactive platform. Dandelions being common in both Susanna's old and new worlds, they connect England and Canada. Readers on both side of the Atlantic, if adventurously inclined, have the possibility to gather the roots and follow her instructions so as to try the coffee for themselves, sharing vicariously in the experimental nature of her experience and transforming text into reality.

Conclusions: food and control

In Moodie's and Traill's accounts of transatlantic learning experiences, food production is one of the success stories, a topic invested with female empowerment: food was under their control. Although throughout most of their food comments the emphasis is on survival and mastering the basics, food sharing is almost inevitably also a matter of enjoyment. This is evident in both sisters' encounter with the most Canadian of Canadian foodstuffs: maple sugar. True to form, Susanna dislikes its taste and texture and makes a meal of her disappointment by describing maple sugar as 'a compound of pork grease and tobacco juice' (Moodie, 1962, p. 114). To sensible Catharine, the glass is half-full:

> In general you see the maple-sugar in large cakes, like bees' wax, close and compact, without showing the crystallization; but it looks more beautiful when the grain is coarse and sparkling, and the sugar is broken in rough masses like sugar-candy. The sugar is rolled or scraped down with a knife for use, as it takes long to dissolve in the tea without this preparation. I superintended the last part of the process, that of boiling the molasses down to sugar; and, considering it was a first attempt, and without any experienced person to direct me, otherwise than the information I obtained from – . I succeeded tolerably well, and produced some sugar of a fine sparkling grain and good colour. Besides the sugar, I made about three gallons of molasses, which proved a great comfort to us, forming a nice ingredient in cakes and an excellent sauce for puddings.
>
> (Traill, 1836, p. 131)

Like all other sobering experiences in her immigrant history, this challenge, too, is turned into an opportunity to demonstrate self-control as a pathway to success. Traill's good breeding leads her to focus on information and on aspects of maple sugar she can praise, i.e. its visual and tactile qualities and the multiple uses to which the product may be put. Her success with maple sugar combines the idea of something quintessentially Canadian with qualities otherwise associated with life in England: comfort and aesthetic pleasure. The text bears witness to her success not only at creating necessaries of daily life but also beautifying extras to render the pioneer hardships more palatable.

Food production and food writing are acts of empowerment and offer ways of connecting past and future, memories and new experiences. Embedded in the project of cultivating land and accommodating oneself to a new continent and level of civilisation, food plays an important role, both for concrete survival and for sustenance of the settlers' sense of self. Food brings people together practically and symbolically, triggers memories and emotions, and food balances the shifting boundaries of the expanding domestic sphere and the wilderness. The act of controlling food keeps the danger of being eaten – practically and metaphorically (by losing one's gentility) – in the minds of the writers and readers, making food into an adventure. Food becomes a relational element between the settler and the land – a balance of mutual consumption is established and a contract of mutual sustainability via sustenance, a relationship of care. By ingesting part of the land, a sense of belonging is generated, as are feelings of entitlement and connection. Ingesting the food grown on the new continent, the settlers learn to internalise Canada and to engage in the hybridising task of building a civilisation which has a firm base in British culture as they know it but constructively encompasses the plants and crops grown in Canada. Heterotopian aspects of food representation function as a shortcut to the predatory ambiguities of colonial appropriation.

In addition to food production, representation and reception, this chapter has made a case for including food contextualisation in considerations of how food and identity constitute one another. Food is intimately connected with place and geography. Both affect where, when and under what conditions food is produced and consumed, represented and received. While few foodstuffs are fit for sending across the Atlantic, writings and drawings about food transcend both time and place: the books survive their authors and continue their transformative work in ever changing contexts. In Moodie's and Traill's texts, food is transcultural: it is connected to a change of location, to travel, emigration and settlement as well as the inevitable transformation of cultures. Being metonymic representatives of their writers, letters about food both recreate and reverse the journey the writers have undergone. In doing so, they continue and correct the hopes that motivated their original move. Tasty food is one of the rare instances of life where utopian expectations and their practical realisation go together, and one might think here of the extended meaning of 'taste' as a general marker of class which is originally food-related. As food production depends on transformation, meals are small steps towards transforming society, one spoonful at a time. While suitably domestic in Victorian terms, the context makes this activity into an empowering transcultural act of discovery and exploration which leaves readers hungry for more.

Notes

1. The Latin original emphasises a connection between handicraft or skill *(ars)* and motherhood; suggestively, it only contains words of feminine grammatical gender. The conventional English translation Moodie alludes to ('invention') serendipitously emphasises the conceptual, imaginative foundation of all craft.
2. Contemporary concerns about food and identity are best epitomised by Njeri Githire's pithy title *The Empire Bites Back* (Githire, 2010). Food in diasporic, transcultural and postcolonial conditions has otherwise been explored in several thematic volumes of journals such as *Moving Worlds, Melus* and *Callalloo,* as well as (Mintz, 1996; Döring et al., 2003).
3. Considerations of the heterotopian potential of food writing are currently missing even in otherwise comprehensive discussions of food culture (Montanari, 2004; Civitello, 2008; Kittler, Sucher, and Nelms, 2012; Counihan and Van Esterik, 2013; Anderson, 2014).
4. Recently, historical dimensions of food as an aspect of cultural tradition are gaining recognition, as documented by the appearance of popular works such as Duncan (2003).
5. Beside the books under consideration here, Moodie wrote a follow-up account that compares the rural Canadian landscapes of her first book with the emerging urban sites, *Life in the Clearings* (1853), as well as in her novel *Flora Lyndsay* (1854). She was a writer of poetry, children's books and novels. In 1985, an edition of her collected letters appeared (Ballstadt et al., 1993). Among her many diverse writings, Catharine Parr Traill addressed various books to prospective immigrants, women as well as children, notably *The Young Emigrants* (1826), *Canadian Crusoes* (1852) and *The Female Emigrant's Guide* (1854).
6. *The History of Mary Prince, a West Indian Slave* (1831) and *Negro Slavery Described by a Negro* (1831).

7 This aesthetic challenges the reader's process of immersing herself in a narrative. Referential aspects of history, geography, biography and authenticity pose unsettling questions about the relationship between fiction and autobiographical modes of writing and different reading conventions, as do attempts to distinguish autobiographical or confessional writings from fiction (Spengemann, 1980). Life writing criticism has moved on from a focus on self and identity as monolithic constructs to analysing modes of relationality (Eakin, 1992, pp. 29–53; Sarkowsky, 2012). Food writing may be seen as relational in that their recipe-like character implies an interactive receiver.

8 Extensive discussions of the genre and terminology of life writing in Canada and beyond attest to its growing importance (e.g. Kadar, 1992b; Eakin, 1999; Gammel, 1999; Jolly, 2001; Kirschstein, 2001; Jensen and Jordan, 2009).

9 Developments of the genre to include female writers are discussed in Hardin (1991); Feng (1998); Moretti (2000).

10 Traill's books on nature and botany span her entire writing life: *The Flower-Basket* (1825), *Sketches from Nature* (1830), *Sketch Book of a Young Naturalist* (1831), *Narratives of Nature* (1831), *Rambles in the Canadian Forest* (1859), *Canadian Wild Flowers* (1869), *Studies of Plant Life in Canada* (1885), and *Pearls and Pebbles; or, Notes of an Old Naturalist* (1894).

11 Catharine Parr Traill writes extensively about her scientific endeavours of classifying plants, labelling and drawing them and inventing names for unknown species (Traill, 1906; Ainley, 1997).

12 Thanks to Traill's renown as a botanist, gardening as part of the settlement mission in the Canadian 'pseudo-wilderness' has received a certain amount of critical attention (Sparrow, 1990; Buss, 1993; Boyd, 2009).

References

Ainley, M. G. (1997). Science in Canada's Backwoods: Catharine Parr Traill. *Natural Eloquence: Women Reinscribe Science*. B. T. Gates and Ann B. Shteir. Madison, WI, University of Wisconsin Press: pp. 79–97.

Amigoni, D., Ed. (2006). *Life Writing and Victorian Culture*. Aldershot, Ashgate.

Anderson, E. N. (2014). *Everyone Eats: Understanding Food and Culture*, 2nd Edition. New York and London, New York University Press.

Appadurai, A. (1996). *Modernity at Large: Cultural Dimensions of Globalization*. Minneapolis, University of Minnesota Press.

Ballstadt, C., E. Hopkins and M. Peterman, Eds. (1993). *Susanna Moodie – Letters of a Lifetime*. Toronto, University of Toronto Press.

Bauman, Z. (1998). *Globalization: The Human Consequences*. Cambridge, Polity.

Beck, U. (1991). *Politik in der Risikogesellschaft*. Frankfurt a.M., Suhrkamp.

Benstock, S. (1988). Authorizing the Autobiographical. *The Private Self. Theory and Practice of Women's Autobiographical Writings*. S. Benstock. Chapel Hill, University of North Carolina Press: pp. 10–33.

Boutelle, A. (1986). Sisters and Survivors: Catherine Parr Traill and Susanna Moodie. *Nineteenth-Century Women Writers of the English-Speaking World*. R. B. Nathan. Westport, CT, Greenwood: pp. 13–18.

Boyd, S. (2009). "'Transplanted into Our Gardens': Susanna Moodie and Catharine Parr Traill." *Essays on Canadian Writing* 84: pp. 35–57.

Buss, H. M. (1993). *Mapping Our Selves: Canadian Women's Autobiography in English*. Montreal, McGill-Queen's University Press.

Campbell, M. B. (2002). Travel Writing and its Theory. *The Cambridge Companion to Travel Writing*. P. Hulme and T. Youngs. Cambridge, Cambridge University Press: pp. 261–273.

Civitello, L. (2008). *Cuisine and Culture: A History of Food and People*, 2nd Edition. Hoboken, NJ, John Wiley.
Conway, J.K. (1998). *When Memory Speaks: Exploring the Art of Autobiography*. New York, Random House.
Cosslett, T., C. Lury and P. Summerfield, Ed. (2000). *Feminism and Autobiography: Texts, Theories, Methods*. London and New York, Routledge.
Counihan, C. and P. Van Esterik, Eds. (2013) *Food and Culture: A Reader*. New York, Routledge.
Deleuze, G. and F. Guattari (1974). *Kafka: Towards a Minor Literature*. Minneapolis and London, University of Minnesota Press.
Döring, T., M. Heide and S. Mühleisen, Eds. (2003). *Eating Culture: The Poetics and Politics of Food*. Heidelberg, Universitätsverlag Winter.
Duncan, D. (2003). *Nothing More Comforting: Canada's Heritage Food*. Toronto, Dundurn.
Eakin, P. J. (1992). *Touching the World: Reference in Autobiography*. Princeton and Oxford, Princeton University Press.
Eakin, P. J. (1999). *How Our Lives Become Stories: Making Selves*. Ithaca, Cornell University Press.
Feng, P.-c. (1998). *The Female Bildungsroman by Toni Morrison and Maxine Hong Kingston*. New York, Peter Lang.
Foucault, M. (1986). "Of Other Spaces." *Diacritics* 16: pp. 22–27.
Freitag, F. (2013). *The Farm Novel in North America: Genre and Nation in the United States, English Canada, and French Canada, 1845–1945*. Rochester, NY, Camden House.
Frye, N. (1995). *The Bush Garden: Essays on the Canadian Imagination with an Introduction by Linda Hutcheon*. Concord, ON, Anansi.
Gammel, I., Ed. (1999). *Confessional Politics: Women's Sexual Self Representations in Life Writing and Popular Media*. Carbondale, IL, Southern Illinois University Press.
Githire, N. (2010). "The Empire Bites Back: Food Politics and the Making of a Nation in Andrea Levy's Works." *Callaloo* 33(3): pp. 857–873.
Gorham, D. (2012). *The Victorian Girl and the Feminine Ideal*. Abingdon and New York, Routledge.
Gusdorf, G. (1980). Conditions and Limits of Autobiography. *Autobiography: Essays Theoretical and Critical*. J. Olney. Princeton, Princeton University Press: pp. 28–48.
Hammill, F. (2003). *Literary Culture and Female Authorship in Canada 1760–2000*. New York, Rodopi.
Hardin, J., Ed. (1991). *Reflection and Action: Essays on the Bildungsroman*. Columbia, SC, University of South Carolina Press.
Huggan, G. and H. Tiffin (2010). *Postcolonial Ecocriticism. Literature, Animals, Environment*. London and New York, Routledge.
Jensen, M. and J. Jordan (2009). *Life Writing, the Spirit of the Age and the State of the Art*. Newcastle, Cambridge Scholars.
Jolly, M., Ed. (2001). *Encyclopedia of Life Writing: Autobiographical and Biographical Forms*. Chicago, Fitzroy Dearborn.
Kadar, M. (1992a). Coming to Terms: Life Writing – From Genre to Critical Practice. *Essays on Life Writing: From Genre to Critical Practice*. M. Kadar. Toronto, Buffalo and London, University of Toronto Press: pp. 3–13.
Kadar, M., Ed. (1992b). *Essays on Life Writing: From Genre to Critical Practice*. Toronto, University of Toronto Press.

Kirschstein, B. H., Ed. (2001). *Life Writing/Writing Lives*. Malabar, FL, Krieger.
Kittler, P.G., K. Sucher and M. Nelms, Eds. (2012). *Food and Culture*. 3rd Edition. Independence, KY, Wadsworth Cengage.
Klepač, T. (2011). "Susanna Moodie's Roughing It in the Bush: A Female Contribution to the Creation of an Imagined Canadian Community." *Central European Journal of Canadian Studies/Revue d'Etudes Canadiennes en Europe Centrale* 7: pp. 65–75.
Klinck, C. F. (1962). Editor's Introduction. *Roughing it in the Bush*. S. Moodie. Toronto, McClelland and Stewart: pp. ix–xiv.
Low, N. (1999). *Global Ethics and Environment*. London and New York, Routledge.
Lynch, C. (2009). Trans-Genre Confusion: What Does Autobiography Think It Is? *Life-Writing: Essays on Autobiography, Biography and Literature*. R. Bradford. Basingstoke, Palgrave Macmillan: pp. 209–218.
Mintz, S. (1996). *Tasting Food, Tasting Freedom*. Boston, Beacon.
Montanari, M. (2004) *Food Is Culture*. New York, Columbia University Press.
Moodie, S. (1962). *Roughing It in the Bush, or Forest Life in Canada*. Toronto, McClelland and Stewart. http://digital.library.upenn.edu/women/moodie/roughing/rough-00.html.
Moretti, F. (2000). *The Way of the World: The Bildungsroman in European Culture*. London, Verso.
Perkin, J. (1995). *Victorian Women*. New York, New York University Press.
Pratt, M. L. (2008a). *Imperial Eyes: Travel Writing and Transculturation*. London and New York, Routledge.
Pratt, M. L. (2008b). Introduction: Criticism in the Contact Zone. *Imperial Eyes: Travel Writing and Transculturation*. London and New York, Routledge: pp. 1–12.
Sarkowsky, K. (2012). "Transcultural Autobiography and the Staging of (Mis)Recognition." *Amerikastudien* 57(4): pp. 627–642.
Saussy, H., Ed. (2006). *Comparative Literature in an Age of Globalization*. Baltimore, MD, Johns Hopkins University Press.
Smith, S. (2001). *Moving Lives: 20th Century Women's Travel Writing*. Minneapolis, University of Minnesota Press.
Sparrow, F. (1990). "'This Place Is Some Kind of a Garden': Clearings in the Bush in the Works of Susanna Moodie, Catharine Parr Traill, Margaret Atwood and Margaret Laurence." *The Journal of Commonwealth Literature* 25(1): pp. 24–41.
Spellman, W. (2002). *The Global Community: Migration and the Making of the Modern World*. Stroud, Sutton.
Spengemann, W. C. (1980). *The Forms of Autobiography: Episodes in the History of a Literary Genre*. New Haven and London, Yale University Press.
Stanford Friedman, S. (1988). Women's Autobiographical Selves: Theory and Practice. *The Private Self: Theory and Practice of Women's Autobiographical Writings*. S. Benstock. Chapel Hill, University of North Carolina Press: pp. 34–62.
Thomas, C. Z. (2009). "'I had never seen such a shed called a house before'. The Discourse of Home in Susanna Moodie's *Roughing it in the Bush*." *Canadian Literature* 203: pp. 105–121.
Thompson, E. (1991). *The Pioneer Woman: A Canadian Character Type*. Montreal, McGill-Queen's University Press.
Thurston, J. (1996). *The Work of Words: The Writing of Susanna Strickland Moodie*. Montreal and Kingston, McGill-Queen's University Press.
Traill, C. P. (1836). *The Backwoods of Canada: Being Letters from the Wife of an Emigrant Officer, Illustrative of the Domestic Economy of British America*. London, Charles Knight.

Traill, C. P. (1906). *Studies of Plant Life in Canada: Wild Flowers, Flowering Shrubs, and Grasses* Toronto, William Briggs.

Whitlock, G. (1985). "The Bush, the Barrack-Yard and the Clearing: 'Colonial Realism' in the Sketches and Stories of Susanna Moodie, C. L. R. James and Henry Lawson." *The Journal of Commonwealth Literature* 20(1): pp. 36–48.

4 Just a happy housewife?

Michelle Obama, food activism and (African-)American womanhood

Katrine Meldgaard Kjær

When Michelle Obama became the forty-fourth First Lady of the United States in 2008, she had made no official statement regarding whether she would pursue a political cause or charity of her own during her time in office. After about a year, however, Michelle Obama would become the very active face of a nationwide campaign battling childhood obesity, thus taking a somewhat sudden turn to becoming very politically active indeed – a circumstance not always easily compatible with the role of First Lady, but which Michelle Obama has nonetheless skillfully maneuvered through, keeping her popularity remarkably intact in the process. In this chapter, I will explore how Michelle Obama has presented her political activity within the realm of food and nutrition to the American public. I will be focusing especially on how this activity has been 'sold' by a constant domestication of both Michelle Obama herself as well as her political motives and motivations, a feat achieved primarily by adhering to certain conventions of gender and gendered rhetoric. I will argue that this domestication and not least personalization has allowed Michelle Obama to inscribe her political activity in a realm of (hetero)normativity that legitimizes her being a politically active First Lady at all. While what follows is a critical analysis of how Obama has pacified her own ambitions and political power in order to remain 'appealing' and relatable, I will also explore how her public gendering of herself and her cause might actually also be read as a progressive step in counteracting the ongoing stereotyping of black and African-American womanhood.

In recent years, much has been written about the Obama presidency. While most of the scholarly attention has been devoted to Barack Obama (Steele, 2008; Pedersen, 2009; Carbado and Gulati, 2013; Jefferies, 2013), Michelle Obama has also received her fair amount of notice, especially in the more popular press (Andersen, 2009; Swarns, 2012). However, although there is a long tradition for thinking about the role of the First Lady and the space for political activism this role allows (Campbell, 1996, 1998; Scharrer and Bissell, 2000; Wertheimer, 2005) and, more recently, about food policies and campaigns that specifically target obesity (LeBesco, 2003; Biltekoff, 2013; Saguy, 2013), relatively little attention has been paid to the intersection of these in Michelle Obama's involvement in *Let's Move!* as a political cause, and even less on how gender and race is a key issue here (Faber, 2009; Guthman, 2011; Puls, 2014). This chapter seeks to expand on the scholarly attention to the intersection of First Ladyship, political food activism, gender and race by asking questions about how Michelle Obama as a First Lady herself negotiates this intersection and what implications this might have.

As cases in point, I will be critically analyzing a selection of Michelle's most publicized rhetoric about her role as both First Lady and face of *Let's Move!* in order to get a sense of the image she is projecting to the American public at large, and, conversely, what discourses she in the process creates around herself and these roles. I will be unpacking this with the help of a framework that is inspired by two different yet intersecting movements within gender studies, namely queer theory and black feminism. Queer theory, pioneered in the 1990s by scholars such as Judith Butler (1990, 1993) and Eve Sedgwick (1990), understands gender as something that is socially constructed through our day-to-day *performances* of gender. Here, gender is not something one is born with, but rather an identity one creates – and, importantly, one that is expected to be created following certain cultural rules and norms of what constitutes 'appropriate' gendered behavior. While agreeing with the premise of gender as being socially constructed, black feminism, pioneered in the same time period by scholars such as bell hooks (1992) and Patricia Hill Collins (1986), argues that there are specific historic and cultural circumstances that influence the performance of gender in an African-American context. They also examine the ways that race and gender always intersect and taint each other and cannot be ignored in a critical investigation of identity construction. In this chapter, I will use this way of thinking about gender and race as a point of departure to unpack the nuances of the performances that make up Michelle Obama's public image.

The chapter will be structured in two parts. The first part of the chapter will focus on the racial and gendered conditions of the role of First Lady and especially how Obama challenges or upsets these conditions. The second part, in turn, will dive into an analysis of how Obama has worked around these challenges and emerged as a major political player in the anti-obesity politics. As such, the chapter will as a whole focus on how food and nutrition can – against the odds – be a legitimate political arena for an African-American First Lady when it is strategically interwoven with 'appropriate' gender performances. The chapter will thus contribute to the anthology by providing a meta-perspective on how the meeting between food and politics can be read as a meeting between private and public, and how identity can be negotiated and public image constructed here.

From 'angry brows' to doting mother: the many faces of Michelle Obama

Today, it is safe to say that Michelle Obama's wide-reaching popularity can be viewed as an asset to her husband. However, when President Obama first started his campaign for office in 2008, Michelle was the object of much more critique than praise. From the very beginning, this critique was especially related to ideas of her supposedly being angry and unpleasant, something that apparently translated into her physicality as well: in 2008 and the start of 2009, for example, Michelle's 'angry' eyebrows and muscular arms were the talk of the town (see, for example, Media Matters for America, 2008). Michelle's then-'angry' image was perfectly illustrated in a now-infamous *The New Yorker* cover from July 2008. Here, Michelle and Barack Obama are pictured as a two-man terrorist operation with Michelle decked out in military-themed gear, wearing army colored camouflage pants, combat boots, a chain of ammunition and what appears to be an AK-47 assault riffle. In addition, she is also portrayed with her hair in the style of an afro, very full lips and her signature 'angry brows'. As noted by Brittney Cooper in an investigation of how Michelle Obama's race challenges still-held conceptions of normative First Ladyship, on this cover 'Michelle Obama's body was used as a canvas to dramatize stereotypes of black female identity' (Cooper, 2010, p. 49).[1] It is important to take a minute to contextualize the representation of Michelle Obama in *The New Yorker*. Indeed, African-American women have a long history of being perceived and represented in America as one of three stereotypical figures: the jezebel, the sapphire and the mammy, corresponding to stock personalities marked by being promiscuous and hypersexual, angry and nagging, and domestic and obedient, respectively (see Johnson, 2003; Givens and Monahan, 2005; Walley-Jean, 2009). On the cover in question, *The New Yorker* clearly represents Michelle as a sapphire, with her militant style, her angry scowl and her big gun just waiting to be shot, thus writing both themselves and her into a long tradition of depicting black women as one-dimensional 'others'. However, when thinking about the root of the anxiety that can explain why Michelle is depicted in this way on, for example, *The New Yorker* cover, we must consider the implications of not only her race, but also her political role as First Lady and, crucially, the combination of these two factors.

The role of the First Lady is a peculiar one. A First Lady is not elected to fulfill her role, has no constitutional powers or duties, and no set of rules or regulations for her to abide by, leaving the seat a largely symbolic and representational one. While the role as First Lady thus consists of a constant negotiation of unwritten rules and expectations, it is clear that the First Lady role is 'a site for traditional femininity' (Anderson, 2005, p. 3). Indeed, this role has been marked for centuries by 'Americans' continuing, deep-seated ambivalence, even hostility, toward power in the hands of women' (Mayo, 1993, p. 15A). All in all, First Ladies are faced with the tackling of 'insuperable obstacles arising out of expectations that they are to represent what we pretend is a single, universally accepted ideal for U.S. womanhood' (Campbell, 1996, p. 191). While a role so undefined casts no obvious bonds on the First Lady, being a site for traditional femininity and American womanhood implies an expectation of a somewhat conservative gendered behavior from the First Lady, without much leeway for deviation: as Hortence Spillers notes, 'the American public apparently does not cotton to behavioral aberration in the presidential wife; Americans seem to prefer that she be seen, not heard' (2009, p. 307). It is with this expectation of conservative gendered behavior that Michelle Obama's powerful profile clashes. She is far from the first First Lady to be as powerful on paper as her husband – in fact, many of the First Ladies have been higher educated than their husbands. However, the most popular First Ladies have done well in downplaying this fact in favor of an image that relied on a more traditional and domestically-oriented feminine ideal – an image that contrasts greatly with Michelle as the militant sapphire described above.

An example of how this ideal of traditional femininity works in practice is the tale of one of the least popular First Ladies ever, Hillary Clinton. Clinton famously spent her first term in the White House in stark and overt opposition to this conservative ideal; in a now infamous TV-interview, for example, she starkly defended her right to her own career, saying, 'I suppose what I could have done is stay at home and baked cookies and had teas, but what I decided to do was to fulfill my profession, which I entered before my husband was in public life' (Making Hillary an Issue, 1992), and later that year, she asserted in a *60 Minutes* interview that she would not be 'some little woman standing by my husband like Tammy Wynette'. Clinton thus openly defied expectations of conservative womanhood. She was accordingly punished consistently in her approval ratings – that is, until she rehabilitated her image after her husband's infidelity scandal by rebranding herself as a stand-by-your-man family woman. After this incident, she stopped talking about her own career and political ambitions, and almost instantly gained popularity, although she never reached Jackie O-like levels of approval. Indeed, the case of Hillary Clinton provides us with some insight into what qualities are apparently not desirable in a First Lady, most notably here being overtly career minded, (economically) independent and openly ambitious.

Interestingly, Michelle Obama's profile is very similar to Hillary Clinton's – they hold the same types of degrees, they were about the same age when they entered the White House, and they both initially profiled themselves as intellectual peers to their presidential husbands. And like Hillary, Michelle got off to a rocky start with the American public, earning her, as Hillary in her time, an unfavorable, hostile and un*womanly* public image, not least on account of the added factor of Michelle's race and the following stereotypical perceptions of black women lending itself to this image; while Hillary was battling 'Americans' continuing, deep-seated ambivalence, even hostility, toward power in the hands of women' (Mayo, 1993, p. 15A), Michelle was battling 'Americans' continuing, deep-seated ambivalence, even hostility, toward power in the hands of women' *and* black people. However, as we shall see, Michelle has actually succeeded in turning her image around to becoming a very popular First Lady in her own right, a feat Hillary Clinton never completely succeeded in achieving.

Rebranding Michelle

Relatively early in Barack Obama's campaign for presidency – that is, from around the publication of *The New Yorker* cover – Michelle was madeover. In contrast to how the last Democratic First Lady Hillary Clinton had been very open about her education, qualifications and ambitions, Michelle Obama suddenly remained mum on the matter. Instead, the public started seeing an increasing amount of pictures of Michelle with her two daughters, and even more pictures of her with Barack

Obama, always smiling and always looking very much in love – and often in the process of receiving a kiss from her husband, the presidential candidate; the 'angry brows', which had by now been re-shaped into a less offensive arch, had been replaced by an avalanche of images portraying Michelle as a warm, maternal figure. At the same time, Michelle re-launched herself as a style-icon, always wearing the most up-to-date and stylish clothes – often from budget-friendly stores like J Crew and the Gap. These images of Michelle as a (hip) mom and wife were supported by the interviews Michelle also starting giving on her own. Here, she would state her unwavering support of her husband and her total commitment to helping him achieve his goals. Her own goals, however, very explicitly took a backseat. In a November 2008 interview with Barbara Walters, for example, Michelle stated that:

> I have never been the kind of person who has defined myself by a career or a job. . . . The truth is, I believe in this man as the president and his vision for the country, and if that meant stepping away from my particular job for a year and a half, or for four or for eight years … then that's a small sacrifice to make. The truth is, and Barack knows this about me, I love kids, and I love my kids and this isn't to say that anyone who's working doesn't.
>
> (*A Barbara Walters Special,* 2008)

With statements like these, Michelle Obama sent out a message that signaled the complete opposite of Clinton's cookie-quote *and* the supposed intentions of the aggressive, angry black woman portrayed on *The New Yorker* cover. The discourse in these interviews is marked by an acknowledgement of the importance of family values and traditional ideals of femininity, executed at one and the same time by confirming her own commitment to domesticity and distancing her from those who do not feel the same commitment. With the statement 'this isn't to say that anyone *who's working* doesn't', Michelle is not only making sure that she is not offending working mothers, she is clearly indicating that *she* is not one of those mothers. This was, of course, later further stressed when she post-election gave herself the title of 'mom-in-chief'. Indeed, six months into the first presidential term, Michelle had still not openly committed to pursuing any specific political arenas or causes. As Michelle's aides told *Washington Post* in April 2009, 'Mrs. Obama hasn't set a firm [political] agenda – besides raising daughters Malia and Sasha' (Bellantoni, 2009). Pair this with how Michelle had also established herself as stylish, contemporary and hip, and an image of woman begins to emerge that is not so easily categorized into the traditional stereotypical black stock figures described earlier.

Both the public and the media seemed be pleased with Michelle's image makeover. Michelle quickly became far more popular than her husband – her approval ratings soared from 43% in 2008 to 72% in mid-2009 (at this time, Barack Obama's approval ratings remained in the mid-fifties range) (Gallup, 2010) – and articles about Michelle Obama being a positive role model of contemporary femininity became commonplace. Michelle and Barack Obama's marriage was also hailed and celebrated, being declared – in an academic production, no less – a marriage that served as 'a reaffirming exemplar of traditional African-American family life and provides an aspirational vision of what African-Americans can attain' (Phillips et al., 2011, pp. 154–155). From the time Barack Obama was elected president, it seemed, the once so threatening Michelle, a supposedly militant and violent angry black woman and part of a marriage plotting to take down the American nation as we knew it, had apparently been de-othered and de-mystified.

Domesticating food activism

But then something interesting happened. In 2010, the *Let's Move!* initiative was launched. *Let's Move!* is a government-funded nationwide anti-obesity initiative that focuses especially on battling childhood obesity. On *Let's Move!*'s website, the official description of the project reads:

> *Let's Move!* is a comprehensive initiative, launched by the First Lady, dedicated to solving the problem of obesity within a generation, so that children born today will grow up healthier and able to pursue their dreams.
>
> (letsmove.org, 2014)

First Ladies choosing a political cause is not unusual; Barbra and Laura Bush both promoted childhood literacy, and Nancy Regan famously urged the youth of the nation to 'Just Say No' to narcotics. What is striking about Michelle's involvement in *Let's Move!*, however, is the discrepancy between the lengths to which she has gone to stress that she is not interested in taking up any other job than taking care of her children, and the political activity she has quite obviously undertaken as the head of *Let's Move!*. An epitomization of this activity came in Spring 2012, where Michelle Obama and *Let's Move!* joined forces with Disney to form *Disney's Magic of Healthy Living*. With this initiative, Disney became the first major American media company to introduce new standards for food advertising and programming targeted towards children. Per the initiative's website, the goal is to have by 2015 'all foods marketed on Disney's television and radio channels required to meet Disney's nutrition guidelines – which align with federal standards to promote fruit and vegetables and limit calories, sugar, sodium and saturated fat' (Duswalt, 2012). What is remarkable about this initiative is that it actually happened – with the free market and the free individual choice being so imbedded in the American consciousness, taking steps to limit this freedom is controversial. Even more controversial is the fact that Michelle Obama as a First Lady is openly positioning herself as being responsible for an initiative that takes these steps; at the press conference announcing the launch of *Disney's Magic of Healthy Living*, Michelle stood on the podium with Disney's CEO Robert Iger and even gave a speech on the importance of this new initiative and how it ties into *Let's Move!*'s overall agenda. Being a central part of a multi-million dollar corporate issue such as Disney's marketing strategies *and* a decision that questions the moral validity of free advertisement and, by extension, the free market, seems to be a far cry from the

Michelle Obama who has so thoroughly stressed her disinterest in working and her adherence to the First Lady's role as a site of a traditional femininity which entails having no political ambitions.

Or is it? As a part of the launching of *Let's Move!,* Michelle Obama introduced the initiative in a two-minute video dedicated to explaining the goals of the initiative as well as the motivation behind her own involvement in it. Strikingly, Michelle Obama's rhetoric has not changed much from the 'happy housewife' narrative put forth in the course of her campaign makeover. In the video, before saying anything about the initiative itself, Michelle recounts the events that prompted her to take an interest in the issues *Let's Move!* is dealing with – this takes just under half the video, so that a full forty-five seconds pass before Michelle beings to talk about the purpose of the initiative. During these forty-five seconds, she tells the story of her past life as a working mother, and the ills her family endured as a result of this:

> Before to coming to the White House, the President and I lived lives like most families; two working parents, busy, trying to maintain some balance. Picking kids up from school, trying to get things done at work. Just too busy, not enough time. And what I found myself doing was probably making up for it in being unable to cook a good meal for my kids. Going to fast food a little more than I'd like, ordering pizza, and I started to see the effects on my family, particularly our kids. It got to the point where our pediatrician basically said 'you may wanna make some changes'. So I started making those changes, short easy changes, but they lead to some really good results. So I wanted to bring the lessons that I learned to the White House.
>
> (The White House, 2010)

Further expanding on her narrative that she does not wish to pursue a career while her husband is in office, Michelle Obama here goes as far as to actively penalize herself for her past life as a *too busy* working parent. The logic put forward in this statement seems to be that because she was too busy trying to lead an active professional life, she did not have time to lead a proper family life and care for her children – caring for her children here understood as cooking home-made, nutritious meals. The working father is rendered completely invisible, as the lack of care in the form of cooking is the fault of Michelle only (communicated by the frequent use of *I*). Luckily, a health professional – importantly, not Michelle herself – intervened and Michelle was able to get her family and children back on track. The *mea culpa* that prefaces the point and purposes of *Let's Move!* thus acts to make it clear that Michelle Obama is not entering into the political realm of food on account of any selfish or self-serving purposes – she is doing it for the nation's children and the families, just like she did for her own.

Likewise, in her speech at the press conference of the launching of *Disney's Magic of Healthy Living,* Michelle Obama kept it personal rather than political, stating that 'just a few years ago, if you had told me or any other mom or dad in

America that our kids wouldn't see a single ad for junk food while they watched their favorite cartoons on a major TV network, we wouldn't have believed you' (The White House, 2012). The First Lady also went on to stress 'as a mom, I know how hard [junk food advertising] makes it to keep our kids healthy' (ibid.). The focus of Michelle Obama's speech was thus on her own experiences as an everyday parent just like 'any other mom or dad in America' and how this new initiative would help her in keeping her kids healthy. Michelle does not speak at length about the common good of the country or about the concrete expected outcomes and implications of the initiative for the population at large. Rather, she speaks on behalf of herself and in relation to her role as a parent. In fact, she even goes so far as to rhetorically separate herself from being any voice of authority by putting herself on the other side of the issue in her statement that she is one of the parents in America that would not have believed *you* if *you* had told *us* that an initiative such as *Disney's Magic of Healthy Living* would have been launched. Here, Michelle Obama is suddenly not a part of the initiative, but merely one of the receivers benefiting from it.

Separating oneself from the voice of power to instead adopt a deeply personal tone, Karlyn Korhs Campbell argues, is a staple of *feminine rhetoric,* that is, the way in which femininity is expected to be enacted or performed rhetorically. Obeying the rules of feminine rhetoric, Campbell writes:

> has meant adopting a personal or self-disclosing tone (signifying nurturance, intimacy, and domesticity) and assuming a feminine persona, e.g., mother or an ungendered persona, e.g., mediator or prophet, while speaking. It has meant preferring anecdotal evidence (reflecting women's experiential learning in contrast to men's expertise), developing ideas inductively (so the audience thinks that it, not this presumptuous woman, drew the conclusions), and appropriating strategies associated with women – such as domestic metaphors, emotional appeals to motherhood, and the like – and avoiding such 'macho' strategies as tough language, confrontation or direct refutation, and any appearance of debating one's opponents.
>
> (Campbell, 1998, p. 5)

By rhetorically pushing herself away from the power by using a personal tone, anecdotal evidence and domestic metaphors, Michelle is actually signaling femininity – a femininity that is much more consistent with her image as 'Mom-in-chief' rather than with her newly found political power. Michelle is thus here not so much a politician or voice of state authority as she is a symbol of American femininity and motherhood – just like every other parent in America. This normalizes Michelle Obama; in her role as a parent, a wife and a concerned citizen, she is above all recognizable:

> Because mothering is a profoundly 'foundational' and 'universal' occupation, Michelle's focus on her family and on the families of others affords her a nuanced rhetorical platform that is nonthreatening, wholesome, and

> comprehensible. It is the female equivalent of 'No Drama Obama'; steady, nurturing, and proactive.
>
> (Kahl, 2009, p. 317)

By inscribing herself in (hetero)normativity, Michelle Obama thus becomes non-threatening and relatable. Unlike Hilary Clinton, who openly questioned the validity of the traditional conservative role of women, and, by extension, First Ladies, Michelle is apparently adhering to the expectations of traditional (passive) femininity that accompanies this role. And while Hillary was punished for her questioning, Michelle is celebrated for her compliance.

As such, Michelle Obama has used her food campaigning/activism as a kind of third space to negotiate the delicate situation she is in as a politically active black First Lady. By making her food activism a space of 'public-in-private and private-in-public' (Sheller and Urry, 2003, p. 108), she can disarm any critique of being too 'public', that is, political, by bringing focus back to how her food activism is actually 'private', that is, domestic. I would argue that the celebration of Michelle Obama is deeply connected with how her adherence to traditional femininity influences how her agenda for engaging in political activity can be perceived overall; by flipping her perspective from being inside the decision-making process to being first and foremost on the outside as a mother working to make her children as healthy as possible, Michelle is not only embodying an acceptable and recognizable gender-based role, she is also legitimizing her very reason for being politically active. By distinguishing herself from a voice of state-power to instead becoming an ordinary woman fighting to set an example and live the best possible life, she embodies yet another culture-specific ideal: the American dream. In her political activity, she remains very much in tune with a collective American imagination marked by individualistic-liberalistic ideals when she constantly stresses that she is involving herself as a part of a striving for the best possible life for her two daughters. She is thus in a way acting individually and out of her own (maternal) interests rather than for the collective good of American society.

In 2012, Michelle Obama moved further into this personal-political realm of food with the release of the first book, *American Grown*. Part cookbook, part gardening book, *American Grown* chronicles the planting and thriving of Michelle's garden in the White House – the first edible garden on the ground of the White House since Eleanor Roosevelt planted her 'victory garden' here during the Second World War. This correlation is intriguing – in her time, Eleanor Roosevelt planted the garden as an encouragement for the people (women) of America to do the same in order to boost morale and food supply around the country in the midst of war. Fast-forward to today, and Michelle Obama is also planting a garden; but this time, the enemy to be fought – obesity – is a domestic rather than foreign. Unlike Roosevelt, Obama is careful not to verbalize or explicitize the garden as a political instrument or statement. Rather, the garden is portrayed as just another case of Michelle Obama being a considerate mother, going about her own business caring for her children and her home. However, as with the launch of *Disney's Magic of Healthy Living,* Michelle goes out of her way to share this personal story with the population at large. Witness, for example, the first words of the introduction to *American Grown:*

> On March 20, 2009, I was like any other hopeful gardener with a pot out on the windowsill or a small plot by the back door. Would it freeze? Would it snow? Would it rain? I had spent two months settling into a new house in a new city. My girls had started a new school; my husband, a new job. My mother had just moved in upstairs. And now I was embarking on something I had never attempted before: starting a garden.
>
> (Obama, 2012, p. 9)

Again, Michelle begins her statement with the phrase 'like any other'. While she was like any other *parent* with *Let's Move!* and *Disney's Magic of Healthy Living,* here, she is like any other *hopeful gardener,* again drawing on the inclusive, relatable and normative. In this drawing on (feminine) normativity, *hopeful* becomes a keyword, as it sets the tone for her newest endeavor as being marked by a certain degree of amateur-ness – something that is also stressed by her mentioning that she had never attempted gardening before. Likewise with the re-occurrence of *just,* which further stresses the extent to which Michelle is not an authority. Once again, this kind of rhetoric works to actively distance Michelle Obama from being any kind of expert on whatever work she is embarking on, instead keeping it strictly personal and anecdotal.

Adding to the 'self-disclosing tone . . . in contrast to men's expertise' is Michelle's characteristic mentioning of her family – her husband, children and, in this case, also her mother – from the very beginning. The domestic/personal setting is thus here acting as *the* catalyst for engaging in the project at hand; in the retrospective offered up in this introduction, planting the garden is part of the transformation Michelle Obama underwent from ordinary citizen to First Lady on equal par with her husband's new job, moving to a new city, her daughters' starting a new school and her mother's moving in upstairs. While these events would not necessarily logically prompt the need to plant a garden, the garden is in this introduction nonetheless framed as the next natural step or progression in this series of events. In the process, the garden thereby becomes a product of not Michelle's own ambitions or aspirations, but rather a natural consequence of her husband's ambitions and aspirations – after all, his new job is the reason she is now suddenly finding herself in a new house in a new city; while her husband works and her children go to school, she tends to her newly planted garden.

Although Michelle Obama enjoys high levels of overall popularity, she has not been immune to critique; an often-voiced criticism of her food activism is that it unfairly interferes with the private lives and choices of ordinary Americans. In a *Good Morning America* (2012) interview to promote *American Grown,* Michelle Obama was asked to comment on the critique voiced of the book specifically and *Let's Move!* more generally as being part of a 'big-government' scheme 'telling me or telling my children what they should eat or should not eat'. Michelle Obama responded to this by again using the domestic as that third space of negotiation, stating that there is no 'one-size-fits-all solution' to the issue of childhood obesity, a fact that she:

> experience[s] [her]self as a mom really trying to figure out in a hectic life how do you take the time to cook a meal, how do you afford it if you don't have accessible, affordable food in your community, how do you get your kids active in a culture where they are spending 7.5 hours a day in front of a screen.

The logic of this statement, however, is flawed: Michelle Obama has repeatedly stated that her most important job is taking care of her children and her family – so important, in fact, that she will not undertake any other job while her husband in in office – and this aspect of her life is by her own previous statements not something she currently lacks time for, and certainly not something she cannot afford. And yet, Michelle, just as in the introductory video to the *Let's Move!* campaign, inserts herself as an example of someone who *in her private life* has in fact struggled with these issues. In the process, she deflects critique by focusing on the struggles that go into striving to be a responsible, good parent – and how *Let's Move!* is really only an initiative that tries to give people – herself included – the tools to work through these struggles.

Furthermore, in this interview, Michelle referrers to childhood obesity as a 'dilemma' rather than the common, more dramatic 'epidemic'. With this choice of words, Michelle not only removes the clinical sound of the obesity to make it instead a more personal issue, she also makes it an issue that is debatable – by using the term 'dilemma', she implies that there are two sides of the given matter to consider, both of which are potentially valid. Obama thus meets some of this critique halfway by granting that she does not have all the answers and by always keeping with the narrative that she is not an authority who has the political legitimacy to tell the American population what they can or cannot do.

A new (African-)American womanhood? Conclusions and perspectives

Despite continued warnings from the press that advocacy could erode her popularity over time (for example from Walsh, 2013), since taking office, Michelle Obama's approval ratings have remained stable – and have consistently showed her as more popular than her husband (Pew, 2013) – and she has become a celebrity in her own right, often being hailed as both a role model and a continuous source of inspiration (Styles, 2012). By not challenging the ideal of traditional womanhood that has come to be expected of First Ladies, it seems Obama has succeeded in avoiding being categorized as a Hillary Clinton-esque type of First Lady, whose blatant ambition and feminist declarations failed to endear her to the public. But perhaps the consequences of Obama's adherence to feminine rhetoric in public are more nuanced, or, at the very least, more multi-faceted than a simple conclusion that because she inscribes herself in normativity, the norm approves. Indeed, the fact that Obama is overwhelmingly celebrated as a positive role model for *black* women in particular deserves consideration. This seems an obvious role for Obama, as she herself is black. Hereafter, the extent to which Michelle's public image has been portrayed as being meaningful for people of especially African-American descent, and the way in which this praise is not necessarily coming only from channels that practice traditional female ideals or conservative femininity is intriguing. Witness, for example, the open letter that R&B superstar Beyoncé penned to Michelle Obama in 2012, where Beyoncé hailed Obama for being 'the ULTIMATE example of a truly strong African American woman. She is a caring mother, a loving wife, while at the same time, she is the FIRST LADY!!! . . . I am *PROUD* to have my daughter grow up in a world where she has people like you to look up to' (Beyonce.com, 2012, emphasis original); or how the feminist website feministing.com declared February 2nd to be the official First Lady Michelle Obama Appreciation Day, arguing that the website on account of being precisely a feminist space needed to be the host for 'tak[ing] a moment to reflect on the wonderful and positive example that has been set by the First African American FLOTUS' (Zerlina, 2012). Or how Jezebel.com, a popular critical general interest women's site, in 2013 called on critics to 'leave Michelle Obama alone', ending their article on the subject matter with the encouraging words, 'Do your thing, Mobama' (West, 2013). So, although I may have painted quite a bleak picture from a feminist point of view about Michelle's own framing of her food activism in this chapter, the way Michelle is presenting herself and her political activity seems, interestingly enough, actually to resonate with groups of people openly declaring themselves feminists or critical of patriarchic structures in society.

When the issue of race and the historical stereotypization of black women discussed previously is taken into account, their celebration is certainly not without merit. While Michelle Obama has not been represented as a jezebel in mainstream media – perhaps on account of the overall respective attitude towards the White House and the Presidency prohibiting such racy imaginations – she has indeed been portrayed as a sapphire, as we saw in *The New Yorker* cover portraying her as violent, aggressive and dangerous, and in the commotion over the angry-ness of her face and eyebrows. However, importantly, she succeeded in breaking away from this stereotype to instead become considered a warm, supportive and loving maternal figure while at the same time also having succeeded in avoiding being stereotyped as a mammy-figure as well, despite the domestic being, as we have seen, a defining element in Michelle's self-portrayal.

Indeed, it is clear that it is Michelle's food activism that provided her with an arena for continuous self-definition as a First Lady, a mother and an American citizen. But evaluating the implications of this self-definition is a not so straightforward. On the one hand, it can certainly be argued that Obama's constant inscribing of herself into traditional femininity and heteronormativity is only confirming an age-old conservative ideal of womanhood that dictates, how one is a wife and mother in a socially acceptable manner and, not least, that power is not to be flaunted if one is born a female; in this interpretation, she is using her status as one of the most visible women in the US, maybe even the world, to promote an image that relies on stereotypes used for so long to keep women 'in their place' and as subordinates to men. However, while I agree that Michelle Obama's public image is fundamentally built on these stereotypes, calling her and her image reactionary would be missing the vital point of race. It must not be forgotten that it has not always been the prerogative of black women to write themselves into some of these roles as wholesome wife and mother; patriarchy mixed with racism has created other straightjacket roles preventing this freedom. So, by profiling herself in relation to wholesome, healthy and heteronormative family ideals, Michelle Obama is actually challenging in an important and significant way the public perception of what black womanhood can entail. By embracing one realm of stereotype (traditional expectation of femininity), she is actually defying another (black female identity as being defined by stereotypical roles such as the jezebel, mammy and sapphire). Food has helped her do this – what better way to 'reclaim a discourse of the (bourgeois) "American dream" and the "common good" of the "American ideal"' (Madison, 2009, p. 324) than through the all-American wholesome family sitting around the all-American wholesome dinner table eating a wholesome all-American dinner (homegrown, naturally)? By choosing to emphasize how she in fact did stay home and bake the (low sugar, low fat) cookies and have the teas that Hillary Clinton denied any interest in, it seems that Michelle Obama is actually negotiating a space for an African-American First Lady to be eligible to be considered the above-discussed 'site for traditional femininity' that represents 'American womanhood'. And that is in itself actually quite a radical move.

Note

1 It is of course important to here note that the editors of *The New Yorker* have firmly and consistently stated that the cover was intended as a parody rather than a literal representation of the Obamas and their respective and collective images. But precisely in being a parody is the cover to some extent a reflection of reality – had these conceptions of the Obamas as generally anti-American and Michelle in particular as angry or threatening in fact not existed or been voiced among the general American public, there would be nothing to make fun of. As such, it makes critical sense to take the parody seriously as an indication of the state of Michelle's image at the time, which was very much marked by a losing battle against 'angry black woman' stereotypes (see also Griffin, 2011).

References

Andersen, C., 2009. *Barack and Michelle: Portrait of an American Marriage.* New York: HarperCollins.

Anderson, K. V., 2005. A Site of American Womanhood. In: M. Wertheimer, ed. *Leading Ladies of the White House. Communication Strategies of Notable Twentieth-Century First Ladies*, Oxford: Lanham Rowman & Littlefield Publishers, Inc., pp. 17–30.

A Barbara Walters Special: Barack and Michelle Obama, 2008. [TV programme] ABC, 25 November 2008.

Bellantoni, C., 2009. Michelle Obama Settling in as a Role Model. *The Washington Times* [online], 10 April. Available at: http://www.washingtontimes.com/news/2009/apr/10/michelle-obama-settling-in-as-a-role-model/?page=all [Accessed 04 December 2014].

Beyonce.com, 2012. *Michelle Obama. Bey-inspired. Beyoncé on our First Lady, Michelle Obama*. Available at: http://www.beyonce.com/news/michelle-obama [Accessed 04 December 2014].

Biltekoff, C., 2013. *Eating Right in America: The Cultural Politics of Food and Health*. Durham: Duke University Press.

Butler, J., 1990. *Gender Trouble – Feminism and the Subversion of Identity*. London: Routledge.

Butler, J., 1993. *Bodies that Matter: On the Discursive Limits of 'Sex'*. London: Routledge.

Campbell, K. K., 1996. The Rhetorical Presidency: A Two Person Career. In: M. J. Medhurst, ed. *Beyond the Rhetorical Presidency*. College Station, TX: Texas A & M University Press, pp. 767–782.

Campbell, K. K., 1998. The Discursive Performance of Femininity: Hating Hillary. *Rhetoric & Public Affairs*, 1(1), pp. 1–20.

Carbado, D. W. and Gulati, M., 2013. *Acting White? Rethinking Race in Post-Racial America*. Oxford: Oxford University Press.

Collins, P. H., 1986. Learning from the Outsider Within: The Sociological Significance of Black Feminist Thought. *Social Problems*, Special Theory Issue 33(6), pp. S14–S32.

Cooper, B., 2010. Ain't I a Lady?: Race Women, Michelle Obama, and the Ever-Expanding Democratic Imagination, *MELUS*, 35(4), pp. 39–57.

Duswalt, M., 2012. *Today the First Lady Announces Disney's Healthy Food Marketing Standards*. Available at: http://www.letsmove.gov/blog/2012/06/05/today-first-lady-announces-disney's-healthy-food-marketing-standards [Accessed 04 December 2014].

Faber, J. M., 2009. ___ Trash in the White House: Michelle Obama, Post-Racism and the Pre-class Politics of Domestic Style. *Communication and Critical Cultural Studies*, 6(3), pp. 39–57.

Gallup, 2010. *Michelle Obama Outshines All Others in Favorability Poll*. Available at: http://www.gallup.com/poll/141524/michelle-obama-outshines-others-favorability-poll.aspx [Accessed 07 December 2014].

Givens, S. M. B. and Monahan, J. L., 2005. Priming Mammies, Jezebels, and Other Controlling Images: An Examination of the Influence of Mediated Stereotypes on Perceptions of an African American Woman. *Media Psychology*, 7(1), pp. 87–106.

Good Morning America, 2012. [TV programme] ABC, 29 May 2012.

Griffin, F. J., 2011. At Last ... ?: Michelle Obama, Beyoncé, Race & History. *Dædalus*, Winter, pp. 131–141.

Guthman, J., 2011. *Weighing In: Obesity, Food Justice, and the Limits of Capitalism*. Oakland, CA: University of California Press.

hooks, b., 1992. *Black Looks: Race and Representation*. Boston: South End Press.

Jeffries, M. P., 2013. *Paint the White House Black: Barack Obama and the Meaning of Race in America*. Stanford, CA: Stanford University Press.

Johnson, E. P., 2003. *Appropriating Blackness: Performance & The Politics of Authenticity*. Durham: Duke University Press.

Kahl, M. L., 2009. First Lady Michelle Obama: Advocate for Strong Families. *Communication and Critical/Cultural Studies*, 6(3), pp. 316–320.

LeBesco, K., 2003. *Revolting Bodies?: The Struggle to Redefine Fat Identity*. Amherst, MA: University of Massachusetts Press.

Letsmove.org, 2014. *Learn The Facts.* Available at: http://www.letsmove.gov/learn-facts/epidemic-childhood-obesity [Accessed 04 December 2014].

Madison, D. S., 2009. Crazy Patriotism and Angry (Post)Black Women. *Communication and Critical/Cultural Studies,* 6(3), pp. 321–326.

Making Hillary an Issue, 1992. [TV programme] ABC *Nightline*, 26 March 1992.

Mayo, E. P., 1993. The Influence and Power of First Ladies. *Chronicle of Higher Education*, 40(4), pp. A52–A54.

Media Matters for America, 2008. *O'Reilly says of Michelle Obama: "She looks like an angry woman."* September 17. Available at: http://mediamatters.org/research/2008/09/17/oreilly-says-of-michelle-obama-she-looks-like-a/145062 [Accessed 05 December 2014].

Obama, M., 2012. *American Grown – The Story of the White House Kitchen Garden and Gardens Across America.* New York: Crown Publishers.

Pedersen, C., 2009. *Obama's America*. Edinburgh: Edinburgh University Press.

Pew Research Center for the People & the Press, 2013. *Obama in Strong Position at Start of Second Term – Support for Compromise Rises, Except Among Republicans.* Available at: http://www.people-press.org/2013/01/17/obama-in-strong-position-at-start-of-second-term/ [Accessed 04 December 2014].

Phillips, C., Brown, L. and Parks, G., 2011. Barack, Michelle and the Complexities of a Black "Love Supreme". In: Parks G. and Hughey, M., eds. *The Obamas and a (Post) Racial America?* Oxford and New York: Oxford University Press, pp. 135–154.

Puls, A. B., 2014. *The First Lady and the American Parent: A Rhetorical Examination of Michelle Obama's Use of Metaphor to Combat Childhood Obesity*. (PhD dissertation). Oregon State University.

Saguy, A., 2013. *What's Wrong with Fat? The War on Obesity and Its Collateral Damage*. Oxford: Oxford University Press.

Scharrer, E. and Bissell, K., 2000. Overcoming Traditional Boundaries: The Role of Political Activity in Media Coverage of First Ladies. *Women & Politics*, 21(1), pp. 55–83.

Sedgwick, E. K., 1990. *Epistemology of the Closet.* Berkeley and Los Angeles: University of California Press.

Sheller, M. and Urry, J., 2003. Mobile Transformations of 'Public' and 'Private' Life. *Theory, Culture & Society,* 20(3), pp. 107–125.

Spillers, H., 2009. Views of the East Wing: On Michelle Obama. *Communication and Critical/Cultural Studies*, 6(3), pp. 307–310.

Steele, S., 2008. *A Bound Man: Why We are Excited about Obama and Why He Can't Win.* New York: Free Press.

Styles, T., 2012. 5 Reasons Michelle Obama Continues to Inspire Us. *huffingtonpost.com* [online], 24 January. Available at: http://www.huffingtonpost.com/tamika-sayles/michelle-obama_b_1225863.html [Accessed 04 December 2014].

Swarns, R., 2012. *American Tapestry: The Story of the Black, White, and Multiracial Ancestors of Michelle Obama.* New York: Amistad/HarperCollins Publishers.

Walley-Jean, J. C., 2009. Debunking the Myth of the "Angry Black Woman": An Exploration of Anger in Young African American Women. *Black Women, Gender & Families*, 3(2), pp. 68–86.

Walsh, K., 2013. Michelle Obama Risks Her Popularity. *usnews.com* [online], 6 June. Available at: http://www.usnews.com/news/blogs/Ken-Walshs-Washington/2013/06/06/michelle-obama-risks-her-popularity [Accessed 04 December 2014].

Wertheimer, M. M., 2005. *Leading Ladies of the White House: Communication Strategies of Notable Twentieth-Century First Ladies*. Oxford: Lanham Rowman & Littlefield Publishers, Inc.

West, L., 2013. Leave Michelle Obama Alone! *Jezebel.com* [online], 24 January. Available at: http://jezebel.com/5978491/leave-michelle-obama-alone [Accessed 04 December 2014].

The White House, 2010. *The First Lady Introduces Let's Move.* Available at: http://www.whitehouse.gov/photos-and-video/video/first-lady-introduces-lets-move [Accessed 04 December 2014].

The White House, 2012. *Remarks by the First Lady at Disney Press Conference.* Available at: http://www.whitehouse.gov/the-press-office/2012/06/05/remarks-first-lady-disney-press-conference [Accessed 04 December 2014].

Zerlina, 2012. First Lady Michelle Obama Appreciation Day! *Feministing.com* [online], 2 February. Available at: http://feministing.com/2012/02/02/first-lady-michelle-obama-appreciation-day/ [Accessed 04 December 2014].

5 'The worst mum in Britain'

Class, gender and caring in the campaigning culinary documentary

Joanne Hollows

In September 2006, Julie Critchlow was dubbed 'the worst mum in Britain' in the UK press (Hattersley, 2006, p. 5). Although not the only woman to receive this title in recent years, Critchlow's 'crime' was to deliver 'junk food' to schoolkids who didn't want to eat the 'healthy food' that their school had started to serve at lunchtimes. Critchlow and her fellow 'sinner ladies' positioned their campaign as a response to the changes to school food brought about by celebrity chef Jamie Oliver's campaign to improve the nutritional content of children's lunchtime meals in his documentary *Jamie's School Dinners* (Perrie, 2006). In the press, the women were likened to killers – they 'effectively shorten the life of these kids' suggested a spokesman for Jamie Oliver (Perrie, 2006) – an image reinforced by references to the ways in which they passed food into the school grounds from a position in the local 'graveyard' (Sims, 2006, p. 29). Although less macabre, the women were variously classified as 'shameful' (Perrie, 2006), as 'lazy' (Stokes, 2006) and as 'rebels without a clue' (White, 2006, p. 4). These judgments were reaffirmed in some readers' letters: for example, *Yorkshire Post* readers described the women as 'ignorant, stupid parents' and 'silly women' who shared 'a single brain cell' and encouraged a 'contempt' for authority that contributed to 'the breakdown of civilized values which protect us from chaos' (Readers' Letters, 2006). If the morality of the 'sinner ladies' hadn't been sufficiently questioned, Jamie Oliver later weighed in by describing them as 'big old scrubbers' on the BBC programme *Top Gear*, reinforcing the ways in which they had been earlier caricatured as 'fat slags' in a cartoon in *The Sun* newspaper (Fox and Smith, 2011, p. 407).

This chapter focuses on a specific format within food TV which emerged in the UK during the past decade – the campaigning culinary documentary (CCD) – to examine how representations of working-class women's food practices are used to not only question their abilities as mothers but also to associate them with a range of moral and social failings. I examine these issues by focusing on the campaigning culinary documentary, a format which blends elements of reality, lifestyle and documentary shows and centres around a problem-solving narrative in which a food personality attempts to solve a food 'crisis'. The chapter builds on my own earlier work on representations of class and gender in *Jamie's Ministry of Food* and *Hugh's Chicken Run* (Hollows and Jones, 2010; Bell and

Hollows, 2011) alongside other recent work which has focused on Jamie Oliver's various forays into the campaigning culinary documentary (Fox and Smith, 2011; Rich, 2011; Warin, 2011; Rousseau, 2012; Gibson and Dempsey, 2013; Piper, 2013).

Much of this research echoes wider debates which highlight how lifestyle and reality television formats have been used to naturalize neoliberal values and to draw distinctions between good and bad consumer-citizens (see, for example, Ouellette and Hay, 2008; Sender and Sullivan, 2008; Couldry, 2010) and draw distinctions between classed identities (see, for example, Haywood and Yar, 2006; Tyler, 2008; Biressi and Nunn, 2013). Frequently, this research is underpinned by the idea that neoliberal rationalities have become part of our culture (Couldry, 2010). Under neoliberalism, the privileging of market competition as the organizing principle for contemporary life is accompanied by attacks on welfare so that 'social service functions are *privatized* through *personal responsibility*' (Lisa Duggan cited in McMurria, 2008, p. 307). Numerous critics have drawn on the work of Foucault to argue that, while neoliberalism represents an attack on statist forms of government, it replaces this with intensified forms of governance which 'encourage people to see themselves as individualized and active subjects responsible for enhancing their own well being' (Larner, 2000, p. 13).

This has led a number of researchers to explore how reality and lifestyle television naturalize neoliberal forms of governmentality. Audiences are asked to do the work traditionally associated with government agencies through 'personal responsibility and self-discipline' (Sender and Sullivan, 2008, p. 574). For example, a number of critics have focused on TV series that deal with health and childcare, areas of life which have frequently been seen as a responsibility of the State. In their discussion of *Honey We're Killing the Kids,* which seeks to makeover parenting practices, Laurie Ouellette and James Hay (2008, p. 476) argue that the show 'put the impetus to succeed in life and health on individuals' and 'teaches personal responsibility, risk-avoidance and choice by diagnosing and rehabilitating cases of "ignorance" and self-neglect'. In this way, the makeover technique used in many of these shows reveals the benefits that accrue to those people who are willing to take responsibility for self-transformation and adopt 'good' lifestyles while pathologizing the 'bad' lifestyles of those who refuse to take responsibility for change.

While the argument I develop here is informed by these debates, I want to focus on why the figure of the working-class mother has been so central in the campaigning culinary documentary and what this can tell us about the relationships between gender, class and food. My analysis is informed by debates about gender and cooking – and, in particular, the work of Marjorie DeVault – which identify how 'feeding work' is primarily a form of caring work through which women produce themselves as 'recognizably womanly' (DeVault, 1991, p. 118). In particular, I focus on the implications of the ways in which working-class women are represented as mothers and the ways in which the ethical dispositions and the domestic knowledge that arises from their class and gendered background are frequently devalued and found lacking in these shows.

The campaigning culinary documentary, food 'crises' and makeovers

In much recent writing about food television, ideas about lifestyle and identity have been seen as central in understanding the significance of new types of TV chef that emerged from the late 1990s onwards. Critics noted a shift away from the instructional and 'how to' approach that characterized many earlier cookery shows (Strange, 1998) to an increasing concern with how cooking and eating were represented as lifestyle practices (Moseley, 2001; Brunsdon, 2003; Hollows, 2003a, Hollows, 2003b; Lewis, 2008). Scholars noted how, in a number of new shows such as Jamie Oliver's *The Naked Chef*, TV chefs were operating as lifestyle experts and cultural intermediaries who promoted the importance of cooking as a lifestyle practice, representing cooking and eating as part of wider projects for constructing the self. The concept of the makeover was frequently used to understand how these shows offered audiences the opportunity to transform and 'improve' their selves and their lifestyles (Moseley, 2000). In the process, some critics noted how the lifestyles and identities that were legitimated and promoted on these shows closely resembled the lifestyles that Pierre Bourdieu (1984) associates with the new middle classes. While some critics suggested that this involved a democratization of new middle-class taste and dispositions (Moseley, 2001), others suggested that lifestyled cookery shows worked to legitimate the tastes and dispositions of the new middle classes and confirm their distinction (Hollows, 2003a; Hollows, 2003b; Lewis, 2007).

A number of critics were also concerned with how these changes were related to a regendering of cookery television (Moseley, 2001). They noted a shift from cookery shows featuring a domestic advisor whose primary address was to women working within the home to primetime cookery television which increasingly offered a vehicle for male restaurant chefs to build brands and demonstrate their culinary skills (Ashley et al., 2004). More generally, a number of critics suggested that as cooking was packaged in primetime entertainment formats, it developed a less specifically gendered mode of address as it was represented as a lifestyle and leisure activity. Some critics also noted how some shows offered the potential for reimagining cooking as a masculine activity (Moseley, 2001; Hollows, 2003a).

Since the early 2000s, a further key shift within food TV has been the diversification of the cooking show into new formats and genres. In this chapter, I want to maintain the focus on questions of class, gender and lifestyle developed in some of the earlier research on cooking but explore them in relation to the campaigning culinary documentary (CCD), a term coined by Hollows and Jones (2010) to identify a new format which emerged on the UK's Channel 4 in which TV chefs launched campaigns to tackle particular food 'crises'. Although this format enables TV food personalities to legitimate their particular approach to food, as I go on to show, it also uses working-class women's food practices as examples of 'bad' lifestyle practices.

Channel 4 introduced the CCD in their primetime '8–9' slot which had become strongly associated with the channel's lifestyle programming (Brunsdon, 2003).

However, by focusing on food campaigns, the format enabled the channel to articulate its star chefs more closely to its public service remit. Much of this was achieved through 'The Big Food Fight', an annual week-long season of shows launched in January 2008 which ran until 2011 and which used CCDs as an anchor for the week. Channel 4's star chefs were positioned not only as lifestyle experts but also as moral entrepreneurs who were able to give focus and leadership to debates about the place of food within national life (Hollows and Jones, 2010). This format would later be deployed elsewhere, most notably in *Jamie's Food Revolution* in the US. Indeed, Jamie Oliver was central to the evolution of the CCD in the UK. As I go on to show, *Jamie's School Dinners* (2005) acted as a blueprint for the new format but Oliver's earlier *Jamie's Kitchen* introduced some of the elements which would characterize the CCD. Focusing on Oliver's attempt to transform a group of unemployed young people into chefs who would become the kitchen brigade in his new restaurant Fifteen (a charitable venture), *Jamie's Kitchen* enabled Oliver to move beyond the lifestyle cookery show and to use his celebrity 'to engage in a more traditional public service role' (Lewis, 2008, p. 62).

A blend of elements from lifestyle and reality television with more 'legitimate' documentary formats would become central to the CCD. As Bell et al. (forthcoming) argue, *Jamie's School Dinners (JSD)* laid down the key characteristics and narrative formula for the CCD. First, *JSD* was set up in response to a perceived 'crisis' (the low nutritional content of British school meals) and centred around a campaign to solve the 'crisis' (by transforming practices in school kitchens and government policy on funding for school dinners). Later CCDs would be motivated by other 'crises' such as poor animal welfare *(Hugh's Chicken Run),* obesity *(Jamie's Ministry of Food)* and the excessive power of supermarkets over food production and consumption *(The People's Supermarket)*. Second, the crisis and campaign in *JSD* provided a framework for a problem-solving narrative in which a food personality must overcome a series of obstacles to realize their ambitions to change food practices for the better. This narrative format provides some of the excitement and tension of the CCD.

Third, *JSD* presented the chef's interventions – and the changes that resulted from them – as the result of the actions of a special and 'inspirational' figure. Jamie Oliver was presented as the only person possessing the passion and knowledge to save children from unhealthy school food. This worked to individualize social change and led to the widespread sanctification of Oliver in the British press where, among other things, he was called a saint (Heathcote Amory, 2005) and 'a latter-day prophet' (Shilling, 2005, p. 2). In contrast, as I go on to explore in more detail, *JSD* represented 'ordinary people' as potential obstacles to the success of Jamie's campaign, a strategy repeated to varying degrees across CCDs more generally. Indeed, in most Channel 4 CCDs, the key 'ordinary people' personifying the obstacles that must be overcome are usually working-class mothers while the inspirational chef has, to date, been entirely male. In what follows I explore how and why this specific gendered (and classed) relationship is central to the CCD.

Fourth, a further key ingredient of the CCD is the way in which the interventions the chef makes to further his campaign frequently rely on his attempt to 'makeover' ordinary people, institutions or industries. Indeed, the attempt to make-over ordinary people frequently provides much of the dramatic conflict – and arguably the entertainment – within the shows. While I go on to show that *JSD* repeated a series of mini-transformations on mothers with bad food practices and school kitchen staff who couldn't cook, in later CCDs this strategy was refined to make more use of the makeover of the 'representative figure' that substitutes for 'the collective experience of working-class life' (Hill, 1986, p. 138). In *Jamie's Ministry of Food,* this is single mother Natasha and, in *The People's Supermarket,* grandmother Josie. In other CCDs, characters who refuse to change operate as a dramatic foil throughout the series: for example, single mother Hayley in *Hugh's Chicken Run.* As I go on to explore, the format therefore relies on a Pygmalion narrative in which some women are redeemed through their willingness to transform and rewarded with a role as the chef's sidekick.

The final characteristic of CCDs is the way they enable food programming and television food personalities to move away from their associations with lifestyle and enable them to address wider social problems. In the case of Jamie Oliver, the CCD has enabled him to trade on the celebrity produced by his investment in lifestyle to recast it as a more serious, more 'national' and more symbolically rich asset (Hollows and Jones, 2010). Nonetheless, the chef's expertise in lifestyle matters – along with the need to transform 'bad' lifestyle choices into 'good' lifestyle choices – remain central to the format.

If 'bad' lifestyle practices around food are sufficient to earn the title of 'the worst mother in Britain', the rest of this chapter focuses on the relationship between the chef and the working-class mother in the CCD to explore how some forms of food practices and knowledge, and some forms of morality and ethics, are judged to be legitimate and superior and others found lacking and in need of transformation. It explores how the key road to redemption for working-class mothers in the CCD is to submit to a transformation while those mothers who resist transformation are offered little reward. In the process, I show how the CCD extends – and offers a particular development within – the lifestyle concerns of contemporary cookery television.

Can't cook, won't cook

In this section I explore how the CCD represents working-class mothers as lacking the skills – and the inclination – to cook or portrays them as empty vessels, devoid of appropriate food knowledge. While my primary focus is on *Jamie's Ministry of Food (JMoF),* this discussion begins with representations of working-class mothers in *Jamie's School Dinners*. I demonstrate how these programmes present working-class mothers' cooking practices and/or food knowledge as in need of reform. I examine how these strategies depend on devaluing feminine knowledge in order to legitimate the right of the professional chef, whose culinary expertise is a product of the public sphere, to legislate on domestic life.

The representation of working-class domestic food practices as a problem is not new. Since the nineteenth century, campaigners have attempted to reform the nutritional content of the working-class diet and these activities have frequently rested on the assumption that working-class culture 'is an impediment to nutrition' (Lupton, 1996, p. 7). More specifically, as Coveney et al. (2012) argue, attempts to regulate and 'improve' working-class cooking skills have focused on teaching 'appropriate' cooking skills to working-class women. Furthermore, they note, debates about cooking skills have become increasingly couched in moral terms and directed towards the working class who are 'required to improve' (Coveney et al., 2012, p. 633).

While domestic cooking practices are not the central focus of *Jamie's School Dinners (JSD),* the importance of healthy school dinners is highlighted through the claim that many children do not receive a healthy meal at home. One of the key impediments to the success of Jamie's campaign for healthy school dinners is children's food tastes: most children initially resist Jamie's healthy offerings because they are shown to have 'bad' taste, a taste for chips and burgers rather than vegetables. Dieticians and doctors reveal that these tastes will eventually kill them. Although the impact of the branding and packaging of convenience and snack foods is noted, these deficient tastes are also shown to have their origins in the kids' homes. This problem is represented by the case of Liam, a working-class boy who is shown to be profoundly neophobic when faced with alien (middle-class) foods which are claimed to be good for him such as organic chicken and asparagus. The self-evident wrongness of Liam's food tastes is reinforced when Jamie tells him that 'Turkey Twizzlers are what idiots eat'. In a quest to change Liam's eating habits, Jamie visits his family and asks his mum to change what they eat at home. Liam's mother is portrayed as an empty vessel who lacks culinary knowledge: she confesses she has never eaten an avocado as she embarks on a trip to the supermarket armed with a shopping list provided by Jamie. Nonetheless, she is shown to be a good mum: she accepts Jamie's prescription for what to eat and his offer to teach her to cook and implements his plan in her home, much to the initial resistance of Liam. Her reward for taking responsibility and being willing to change is that Liam himself is allegedly transformed as he is weaned off a sugar-heavy diet. When Jamie returns, we hear that Liam has become 'calmer' and 'more loving'. As the voiceover makes clear, if Jamie is to realize his aims, 'its not just a question of changing the children, Jamie has also got to change the parents'.

Should we be in any doubt about the difference between 'good' and 'bad' parental cooking practices, the series returns us to scenes of Jamie cooking for his own children. Indeed, 'good' parenting is established a few minutes into episode one when we see Jamie educating his young daughter about the various vegetables he is preparing and we later learn that cooking is one activity guaranteed to cheer her up. His own domestic practices are used to transform the children who are most resistant to his new healthy school menus. Jamie educates the kids about different vegetables and gets them cooking as a route to getting them to eat unfamiliar foods. 'I think its getting them involved, giving them ownership' he says, with the implication that their parents are failing to do this. Indeed, aside from a couple of incidental male dinner 'ladies' in the final episode, Jamie is the only man who cooks for children throughout the show. *JSD* simultaneously naturalizes the idea that caring for children's health through feeding work is female labour while demonstrating that Jamie's culinary practices in both public and private spheres are infinitely superior to those of mothers. This is even reaffirmed in his own home where, in his wife's presence, he criticizes her shopping skills to camera, claiming that her selection of 'unseasonal' products is 'driving me mad'.

Central to the narrative and drama of *JSD* is Jamie's relationship with Nora, a school 'dinner lady' at Kidbrooke School, which acts as the prototype for transforming school dinners. While Nora is largely presented in a positive light and becomes a willing advocate for change alongside Jamie, the programme nonetheless undermines her cooking skills by comparing her to the professional chefs in Oliver's restaurant Fifteen. Although dinner ladies are partly shown to have been deskilled through their requirement to reheat processed food and Nora is represented as highly competent in running her kitchen, Nora's culinary skills are called into question. Jamie sends her to spend a day at Fifteen with his head chef Arthur Potts-Dawson, who begins by teaching her 'knife skills'. Potts-Dawson jokes at her expense that 'it's funny but most chefs use knives', to which Nora replies, 'I'm a cook not a chef'. However, the programme suggests that being a cook is not sufficient, reproducing ideas that masculine culinary practice is superior to feminine domestic cookery (Mennell, 1996). Indeed, at the end of her day at the restaurant, this distinction is reinforced when a young chef says approvingly that Nora has the potential to be a chef not just a cook.

If Potts-Dawson is tasked with 'inspiring . . . passion for food in Nora', then Nora in turn is tasked with inspiring this passion in other dinner ladies in order to roll out Jamie's reforms across the school district. As Nora puts it: 'Once people see somebody enthusiastic, anyone who actually wants to change it around will go with you one hundred per cent'. This not only partly responsibilizes the dinner ladies for the success of Jamie's mission – all they need is the *desire* for change – but it also, like a number of reality shows, reproduces the importance of 'passion' as an obligation within the neoliberal workplace (Couldry, 2010). While we witness Jamie's commitment to the project resulting in excessively long working days with little chance for leisure or family time, after one day of cooking Jamie's healthy school menu, dinner lady Lesley complains of being 'exhausted' and on a 'downner'. She says:

> I'm not willing to kill myself to do it. . . . This is not my life this job. It might be other people's lives like Nora. . . . It's not my life. It's just a tiny little bit of my life. I work to live. I don't live to work.

Although Nora is visibly upset by the extent to which the dinner ladies are stressed by the changes and Jamie asks the local council to extend their paid working hours, *JSD* also invites the audience to judge Lesley for her lack of 'passion'. If passion is an obligation in the neoliberal workplace because it 'erases contradictions and legitimates the extended appropriation of the worker's time' (Couldry, 2010, p. 76), then *JSD* legitimates this by demonstrating how 'good' citizens like Jamie and

Nora put others' needs before their lives because they have the passion. Furthermore, by adopting the view that she simply 'work[s] to live', Lesley is also judged for her lack of care for the children she feeds, an attribute that women are meant to bring 'naturally' to their practices. Similar strategies were reproduced in *Jamie Oliver's Food Revolution* in the US (Slocum et al., 2011).

Therefore, *JSD* delegitimates working-class women's culinary skills in both public and private spaces, showing them to be at best 'mere' cooks and, more usually, to be culinary incompetents with poor food knowledge and only sufficient skill to reheat processed food. The show suggests that good mums are like Liam's mum, empty vessels who willingly absorb Jamie's food knowledge and agree to self-transformation. A visual fascination with the dinner ladies' cigarettes – and a preoccupation with their demand for cigarette breaks – links their bad culinary practices with other dangerous and illegitimate lifestyle practices. Therefore, while *JSD* demands that the working-class women featured in the show should cook for children – as mothers or as their dinner ladies acting in culinary loco parentis – their practices are simultaneously delegitimated because they are represented as lacking the level of skill, knowledge and care that characterizes Jamie's practice. Indeed, should anyone be in doubt about the illegitimacy of working-class parents' food practices, Jamie makes it clear in the follow-up show *Return to Jamie's School Dinners*. Claiming that he's had enough of 'being PC about parents', Jamie states:

> If you're giving very young kids bottles and bottles of fizzy drink, you're a fucking arsehole, you're a tosser. If you're giving them bags of fucking shitty sweets at that very young age, you're an idiot. . . . If you never cook a hot meal, sort it out, do it once a week please. . . . I'm here because I truly care. . . . If they truly care . . . they've got to take control.

As I go on to show, an ideal type of this kind of 'idiot' parent is offered up in *Jamie's Ministry of Food*. In the opening sentence of episode one, the voiceover tells us that 'Kaya has never had a home-cooked meal', and shortly after, Kaya's mother Natasha reveals her fridge to be full of confectionary and fizzy drinks.

However, this isn't the only direct link between the intertexts of *JSD* and *Jamie's Ministry of Food (JMoF)*. 'The worst mother in Britain' Julie Critchlow – linked to *JSD* through media coverage of her resistance to Oliver's campaign – is used as our point of entry into *JMoF*. Jamie goes to visit Julie Critchlow, telling us: 'She represents everything that's wrong with our relationship to food in this country'. However, Julie is partially redeemed because it is revealed that she can and does cook, a point later reinforced when she cooks him a roast dinner which, in the UK, operates as the 'proper meal *par excellence*' (Charles and Kerr, 1988, pp. 18–19). Furthermore, the reason that Julie can cook is because she was taught to cook by her mother. Mothers' abilities as cooks – and their responsibility for 'passing on' culinary skills and knowledge – becomes a key focus of the series.

Jamie's campaign in *JMoF* appears to be motivated by the aim to 'get Britain healthy again': as the book of the series claims, 'we have a . . . war on our hands

. . . over the epidemic of bad health and the rise of obesity' (Oliver, 2008, p. 9). However, as Hollows and Jones (2010) observe, there is a slippage between a campaign teaching people to eat more healthily and one teaching them culinary skills, as if good cooking automatically produced leaner citizens. The series therefore rests on the assumption that healthiness is a product of home cooking, despite the fact that a significant number of recipes in the accompanying cookbook do not conform to dominant dietary recommendations. In this way, *JMoF* reproduces contemporary discourses in which childhood obesity is linked to cooking skills and, in particular, women's lack of culinary skill and their dependence on convenience foods (Coveney et al., 2012). This also demonstrates the 'normative power' of a 'cooking-from-scratch discourse' in which home cooking is seen as an antidote to unhealthy eating (Halkier, 2010, p. 105).

Jamie's mission in *JMoF* is structured around his plans to find a new way to 'pass on' the cooking skills which are no longer being transmitted in the home. His plan to get people cooking is to get people to pass on recipes using traditional forms of face-to-face communication so that culinary skills could be passed on 'like spreading gossip' or 'like a dose of the clap'. Jamie's campaign to 'pass it on', 'is essentially a modern-day version of the way people used to pass recipes down through the generations, when they weren't all at work' (Oliver, 2008, p. 12). Should we be in any doubt who *they* refers to, Oliver would later claim on the TV show *Friday Night With Jonathan Ross* that the decline in cooking skills was because 'our girls have been sent out to work for the last 40 years' (cited in Fox and Smith, 2011, p. 405). Having diagnosed contemporary women's contribution to the problem, his plan is therefore to recapture an idealized notion of a feminine domestic tradition that lies outside modernity, in which 'repetition, understood as ritual, provides a connection to ancestry and tradition' (Felski, 2000, p. 83), in this case a lost national-domestic culinary culture. This emphasis on women's maternal responsibility for feeding work and their symbolic role 'as markers of the nation's moral values' (Skeggs, 2005, p. 968) is reinforced throughout the series: the primary female characters (including Jamie's wife Jools) are defined by their role as mothers ('single mum Natasha') while the key male figures are defined by their occupation ('Mick the Miner').

Natasha Whiteman is also the central 'non-cook' who is used to articulate concerns about class, gender and mothering. As Hollows and Jones (2010) have argued, the series uses aesthetic and narrative strategies informed by earlier film representations of the British working-class, making use of visual strategies to establish a distant viewing position and 'fasten' people in their environment (Colls and Dodd, 1985, p. 29). Furthermore, *JMoF* deploys Natasha as a 'representative' figure of her class (Hill, 1986, p. 138) who is used to suggest widespread problems with working-class food practices. Imogen Tyler (2011, p. 210) notes how such strategies mobilize 'a conjunction of signs, bodies and landscape which compose a familiar assemblage of classed and gendered values', what Bev Skeggs and Helen Wood call a 'moral subject semiotics' (cited in Tyler, 2011, p. 210). In a much discussed scene in the opening episode, we are introduced to Natasha who confesses her inability to cook and is positioned by her 'inappropriate' consumption practices. Natasha is not represented in terms of her 'inability to consume' but in terms of consumption practices marked as '*aesthetically* impoverished' (Hayward and Yar, 2006, p. 14). Therefore her 'inappropriate' feeding practices, represented through a fridge full of confectionary and her children eating kebabs out of Styrofoam containers on the floor, are juxtaposed with shots of her eight-ring cooker (which she confesses she can't use) and a supersized television screen. The significance of this 'inappropriate' consumption was not lost on viewers who commented on blogs (Hollows and Jones, 2010), findings backed up by audience research on the series (Jackson et al., 2013; Piper, 2013). Natasha's 'poor' feeding practices are shown to be a result of 'cultural' rather than economic poverty, a point reinforced in commentary by Jamie Oliver himself in which he claimed:

> I'm not judgemental but I have spent a lot of time in poor communities, and I find it quite hard to talk about modern day poverty. You might remember that scene . . . with the mum and the kid eating chips and cheese out of Styrofoam containers, and behind them is a massive fucking TV. It just didn't weigh up.
>
> (Deans, 2013)

The ways in which the significance of objects is constructed through 'visual shots and sequences that incite negative moral judgement' (Tyler, 2011, p. 215), enables these images to have resonance beyond the actual show (Jackson et al., 2013) and conjure up specific associations.

A key narrative strand of the series concerns Natasha's rehabilitation. *JMoF* draws on conventions from lifestyle and reality series which teach 'personal responsibility, risk-avoidance and choice by diagnosing and rehabilitating cases of "ignorance" and self-neglect, and allowing the viewer at home to identify as normal in comparison' (Ouellette and Hay, 2008, p. 476). From the outset, Natasha confesses her guilt in relation to her child, saying she has done things that are 'easier for me than for her'. In this series, Natasha not only takes on the responsibility of learning to cook and teaching others to do so but these activities are also part of her wider responsibilization as she becomes an advocate for Jamie's mission and enrolls in a culinary training course which promises to move her away from dependence on state benefits. Indeed, similar points about the ways in which CCDs can promote forms of neoliberal governmentality have been made in studies of *Jamie's Food Revolution* in the US, in which participants are constructed as 'ignorant' but amenable to 'correction' (Rich, 2011; see also Slocum et al., 2011; Warin, 2011; Gibson and Dempsey, 2013).

However, the use of Natasha as a 'representative' figure does not simply work to identify irresponsible feeding work with the working class, but also to identify working-class mothers – rather than parents in general – as those who must be responsibilized. As we have already seen in the press coverage of Julie Critchlow, 'bad' feeding work is seen to symbolize bad mothering. Such responses were evident in readers' comments on the *Daily Mail* newspaper's website in response to Natasha, where her practices were likened to 'child cruelty [*sic*]' and 'a form of child abuse' (Hollows and Jones, 2010, pp. 315–16). The vilification of working-class mothers in particular can be partly understood with reference to empirical studies of gender and cooking which demonstrate how women's work in 'feeding the family' is primarily a form of caring work, the 'undefined, unacknowledged activity central to women's identity' (DeVault, 1991, p. 4). For DeVault, 'feeding work' not only comprises cooking but a range of other tasks such as planning, scheduling and shopping which are necessary to feed families and which involve a vast amount of mental, manual and emotional labour that remains largely invisible. For DeVault (1991, p. 118), it:

> is not just that women do more of the work of feeding, but also that feeding work has become one of the primary ways in which women 'do' gender. . . . By feeding the family, a woman conducts herself as recognizably womanly.

Given these associations between women – and, in particular, mothers – and food and caring, Natasha's identification with convenience rather than care (Warde, 1997) not only marks her as a failed cook but as failing to be 'appropriately' feminine.

JMoF, therefore, not only contributes to a wider climate in which working-class consumption practices are pathologized and deemed to be a threat to the nation's 'health' (Hayward and Yar, 2006; Tyler, 2008; Biressi and Nunn, 2013), but focuses on the need to reform working-class mothers' practices if the future health of the nation is to be secured. By portraying working-class mothers as insufficiently caring, they are also portrayed as failed women and bad mothers. However, while these representations of working-class mothers 'enable audiences to experience and produce class differences: to affirm, "*I am not that*" ' (Tyler and Bennett, 2010), they also enable male food personalities such as Jamie to distinguish themselves through their ability to improve domestic practice and their capacity to care. The significance of this identification of culinary professionals as domestic experts and extraordinarily caring is explored further in the next section when I go on to explore these issues in relation to ethical consumption.

Caring, ethics and food consumption

In this section, I want to examine how working-class women are represented as 'unethical' consumers. While the responsibility for ethical consumption has become part of women's feeding work (Cairns et al., 2013), I demonstrate how CCDs disavow the significance of the 'ordinarily ethical' dimensions of everyday cooking practices in favour of the '*performative* practice associated with being an ethical consumer' (Barnett et al., 2005, p. 41). In the process, I highlight how class and gender impact on the extent to which caring can operate as a form of capital. The section focuses on *Hugh's Chicken Run (HCR),* broadcast in 2008 as part of Channel 4's season of food programming, The Big Food Fight, in which TV chef Hugh Fearnley-Whittingstall campaigned for the ethical production and consumption of chicken by highlighting the poor welfare of chickens that were intensively produced. I extend my points in relation to *The People's Supermarket* (*TPS*, 2011), which documented chef Arthur Potts-Dawson's attempts to establish a more ethically responsible alternative to the major supermarket chains. Both series rely on attempts to makeover food consumers' habits as a key part of their campaign, and in both series, a working-class woman is the chef's key adversary.

While *JMoF* places much of the blame on individual food consumption practices for the 'obesity crisis', *HCR* highlights the major supermarkets' responsibility for the extent of intensively reared 'cheap chicken' production as well consumers. Nonetheless, as Bell and Hollows (2011) show, a central narrative of the series focuses on an attempt to transform the chicken-buying practices of working-class residents of Axminster's Millway Estate, located in what Hugh calls 'the tough end of town' where people are 'struggling with their food budget'. In order to persuade the residents to abandon their dependence on cheap supermarket chicken, Hugh teaches them about producing their own food on a local allotment and confronts them with the grim reality of the lives of intensively reared birds. The series climaxed in a 'free-range week' in which all the residents of Axminster were asked to buy only free-range birds. In the process, like the green lifestyle television discussed by Tania Lewis (2008, p. 238), in *HCR* 'regulating one's consumption and embracing the necessary inconveniences of green modes of living are offered up as middle-class virtues to which we should all aspire'. As Lewis observes, 'ethical modes of distinction are increasingly associated with social distinction' yet the resources needed to engage in ethical consumption practices are 'unevenly distributed across lines of class, gender, race and ethnicity' (Barnett et al., 2005, p. 41).

However, while the rest of the Millway residents are transformed into 'ethical consumers', a central narrative of the series focuses on single mother Hayley's resistance to being madeover and, implicitly, positions her as an 'unethical consumer'. While Hayley works tirelessly on the allotment, she refuses to demonstrate the 'appropriate' emotional dispositions on witnessing the grim reality of intensive chicken production. While we witness Hugh weeping over – and caring about – the plight of cheap chickens, Hayley remains unmoved, informing Hugh that these chickens are 'for people's pockets like mine. You can afford the free range, I can't'. As Bell and Hollows (2011) argue, Hayley constantly reiterates that her primary ethical responsibility is to care for her family through budgeting wisely. Here the 'ethics' of ethical consumption come into conflict with the 'ordinarily ethical' dimensions of everyday domestic practices (Barnett et al., 2005, p. 28) oriented around values such as thrift which enable Hayley to be recognized as a 'good' and responsible mother (DeVault, 1991; Miller, 1998). The series reflects on these issues to some extent when, towards the end of the final episode, Hugh discovers Hayley buying cheap chicken during his 'free-range week'. Hayley recognizes that she is positioned as an 'unethical consumer' and states: 'Don't look at me like that' and 'This is all I can afford at the moment'. While Hugh temporarily reflects on the limits of ethical consumption – 'Back to reality. Mums like Hayley, tough budgets, kids to feed, two [chickens] for a fiver, what are you going to do?' – the audience along with Hugh is nonetheless left looking at Hayley 'like that', as someone who does not care enough about the plight of cheap chickens, as the show turns back to celebrate those who have their ethical priorities 'right' (Bell and Hollows, 2011).

Hayley's inability to generate any form of profit from the ethical dimensions of her feeding work are not only a result of the ways in which caring work has been undervalued because it is associated with what women do 'naturally', but also because publically recognized forms of ethical consumption (for example, shopping for organic produce) are increasingly being seen as part of the feeding work through which women can construct themselves as 'good' mothers. In their study of mothers' ethical consumption practices in Canada, Cairns et al. (2013, pp. 97–8) note the 'gendering of ethical food discourse' which produces the idea that 'mothers are deemed individually responsible for producing a healthy child *and* a healthy planet'. This responsibility is individualized so that women who cannot afford to take part in what have become required ethical practices are positioned in terms of 'personal deficiency, rather than a product of structural inequality' (Cairns et al., 2013, p. 111).

If Hayley's unwillingness to take responsibility for the planet as well as her children marks her as deficient, then some similar issues emerge in another CCD, *The People's Supermarket*. While less centrally concerned with making over consumer practices than many other CCDs, central conflicts within the show concern people's unwillingness to use the cooperative because it is too expensive and the related issue of its policy on stock. Unlike the other CCDs discussed here, the location of *TPS* in central London also means that it features both middle-class and working-class local residents. While chef Arthur Potts-Dawson is concerned that the shop addresses the needs of all members of the local community, the prices in the shop are high because he lacks the buying power of the major supermarkets and because he wants to adhere to ethical standards in relation to local produce, environmental sustainability and non-exploitative relations with suppliers. In a number of scenes across the earlier episodes we see the supermarket's middle-class members perform ethical distinction in their requests for organic and locally sourced fresh produce while working-class women are frequently left defensively requesting 'plain basic stuff' on which they can feed their families for a reasonable price. The 'ethical' dimensions of these women's tastes remain largely invisible while, despite sympathetic responses from Arthur, the show works to reinforce middle-class forms of ethical consumption which demonstrate a responsibility to both family and planet as normative.

Furthermore, both Hugh Fearnley-Whittingstall and Arthur Potts-Dawson are, like Jamie Oliver, represented in terms of their exceptional levels of compassion, passion and caring, qualities demonstrated by visible displays of emotional intensity. If 'compassion is now part of the job description of the contemporary celebrity' (Goodman, 2011), these shows work to brand their stars in terms of their capacity to care and enable them to use their moral and ethical investments to generate profits in distinction as they develop and enhance their brand image as moral entrepreneurs (Hollows and Jones, 2010). The emphasis on passion and emotional investment also legitimates the organization of work under neoliberalism. As Nick Couldry (2010, pp. 75–6) argues, 'Passion becomes a necessity in the neoliberal workplace because its work of denial erases contradictions and legitimates the extended appropriation of the worker's time'. If a constant refrain in public responses to *JMoF* was that 'at least he's doing something' (Hollows and Jones, 2010), then these series more generally suggest that caring about – and caring through – feeding work is not simply a duty but also requires passionate devotion. However, given that women's feeding work within the family is usually unpaid (and the feeding work of school dinner ladies is usually performed for low wages), as Slocum et al. (2011, p. 185) observe in relation to *Jamie's Food Revolution* in the US, 'Characterizing Oliver's methods of cooking as the embodiment of love and care suggests that women's inadequately compensated labour is justified by being done out of love rather than for a wage'.

Finally, the ethical dispositions and caring practices of the male protagonists in these shows are legitimated over the more 'ordinarily ethical' caring practices involved in everyday domestic feeding work. In her discussion of Joan Tronto's work, Elizabeth Silva (2007, p. 143) distinguishes between 'care as practice', which 'refers to the material accomplishment of tasks and activities', and 'care as disposition', which 'refers to the emotional investment in caring'. These series devalue the ordinary work of 'caring for' while reproducing the idea that it is a feminine responsibility, and instead privilege 'caring about', which Tronto (1989, p. 174) characterizes as traditionally a more masculine disposition. As we have seen, 'caring about' chicken welfare or the 'obesity crisis' can be used by TV chefs to generate a sense of distinction and the authority to legislate on the proper conduct of domestic feeding work. However, working-class women are responsibilized for both 'care for' and 'care about', practices which are frequently intertwined but which can also produce contradictions. In *HCR* and *TPS*, forms of ethical consumption are emphasized as forms of 'caring about' that are directed towards the public sphere which, for working-class women, come into conflict with the economic constraints of the everyday practices of 'caring for'.

Conclusion

If primetime cookery shows have contributed to the lifestylization of cooking in ways that privilege new middle-class dispositions and open up the potential of masculinizing domestic cookery, campaigning culinary documentaries work to highlight 'bad' feeding practices and associate them with working-class women. While this can be understood as a continuation of a longer history in which working-class food practices in the UK have been seen as culturally as well as nutritionally deficient, these contemporary shows articulate these concerns within neoliberal modes of governmentality in which individuals must be responsibilized for both their own – and the wider social – good. Furthermore, as Fox and Smith (2011, p. 411) observe in their findings on *Jamie's School Dinners*, 'the middle class are not subject to the same level of media surveillance as the perceived failures of the lower orders'.

The narratives of these shows are predicated on an unequal relationship between male food personalities and working-class women. While these series (to differing degrees) allow space for these women to question the terms of the chefs' campaigns, their narrative redemption is dependent on their willingness to change their feeding work and their lifestyles. Although most of these series include men within the remit of their campaigns, it is working-class mothers who act as representative figures – and as exemplars to the audience – of those whose practices are most in need of transformation. While these shows do not devalue domestic life, women's expertise in domestic feeding work – and the everyday practices of 'caring for' – are shown to lack value compared to male celebrity chefs who use their expertise to improve domestic life and are distinguished by their exceptional ability to 'care about' bigger issues. While this could be read as progressive because it undermines assumptions that women are more naturally suited to the caring work of feeding others, the Jamie Oliver-led CCDs position women as responsible for feeding work but as failing to carry it out responsibly.

References

Ashley, B., J. Hollows, S. Jones, and B. Taylor, 2004. *Food and Cultural Studies*. London: Routledge.

Barnett, C., P. Cloke, N. Clarke, and A. Malpass, 2005. Consuming Ethics: Articulating the Subjects and Spaces of Ethical Consumption. *Antipode*, 37(1), pp. 23–45.

Bell, D. and J. Hollows, 2011. From *River Cottage* to *Chicken Run*: Hugh Fearnley-Whittingstall and the Class Politics of Ethical Consumption. *Celebrity Studies*, 2, pp. 178–91.

Bell, D., J. Hollows, and S. Jones, forthcoming. 'Campaigning Culinary Documentaries and the Responsibilization of Food Crises', *Geoforum*.

Biressi, A. and H. Nunn, 2013. *Class and Contemporary British Culture*. Basingstoke: Palgrave Macmillan.

Bourdieu, P., 1984. *Distinction: A Social Critique of the Judgement of Taste*. London: Routledge.

Brunsdon, C., 2003. Lifestyling Britain: The 8–9 Slot on British Television. *International Journal of Cultural Studies*, 6, pp. 5–23.

Cairns, K., J. Johnston, and N. MacKendrick, 2013. Feeding the 'Organic Child': Mothering through Ethical Consumption. *Journal of Consumer Culture*, 13(2), pp. 97–118.

Charles, N. and M. Kerr, 1988. *Women, Food and Families*. Manchester: Manchester University Press.

Colls, R. and P. Dodd, 1985. Representing the Nation: British Documentary Film, 1930–45. *Screen*, 26(1), pp. 21–33.

Couldry, N., 2010. *Why Voice Matters: Culture and Politics After Neoliberalism*. London: Sage.

Coveney, J., A. Begley, and D. Gallegos, 2012. 'Savoir Fare': Are Cooking Skills a New Morality? *Australian Journal of Adult Learning*, 52(3), pp. 617–42.

Deans, J., 2013. Jamie Oliver Bemoans Chips, Cheese and Giant TVs of Modern-day Poverty. *Guardian*, [online] 27 August. Available at: <http://www.theguardian.com/lifeandstyle/2013/aug/27/jamie-oliver-chips-cheese-modern-day-poverty> [Accessed 5 December 2013].

DeVault, M., 1991. *Feeding the Family: The Social Organization of Caring as Gendered Work*. Chicago: University of Chicago Press.

Felski, R., 2000. *Doing Time: Feminist Theory and Postmodern Culture*. New York: New York University Press.

Fox, R. and G., Smith, 2011. Sinner Ladies and the Gospel of Good Taste: Geographies of Food, Class and Care. *Health and Place*, 17, pp. 403–412.

Gibson, K.E. and S.E. Dempsey, 2013. Make Good Choices, Kid: Biopolitics of Children's Bodies and School Lunch Reform in Jamie Oliver's *Food Revolution. Children's Geographies*, 33(1), pp. 44–58.

Goodman, M., 2011. Celebritus Politicus and Neo-liberal Sustainabilities. *Environment, Politics and Development Working Papers Series*, King's College London, No. 28.

Halkier, B., 2010. *Consumption Challenged: Food in Medialised Everyday Lives*. Farnham: Ashgate.

Hattersley, G., 2006. We Know What Food the Kids Like, and It's Not Polenta. *Sunday Times*, 24 September, p. 5.

Hayward, K. and M. Yar, 2006. The 'Chav' Phenomenon: Consumption, Media and the Construction of a new Underclass. *Crime Media Culture*, 2, pp. 9–28.

Heathcote Amory, E., 2005. Why We Should All Back the Naked Chef's Crusade to Stop Feeding Our Children Junk Food. *Daily Mail*, 11 March, p. 14.

Hill, J., 1986. *Sex, Class and Realism: British Cinema, 1956–63*. London: BFI.

Hollows, J., 2003a. Oliver's Twist: Leisure, Labour and Domestic Masculinity. *International Journal of Cultural Studies*, 6(2), pp. 229–48.

Hollows, J., 2003b. Feeling Like a Domestic Goddess: Postfeminism and Cooking. *European Journal of Cultural Studies*, 6(2), pp.179–202.

Hollows, J. and S. Jones, 2010. 'At Least He's Doing Something': Moral Entrepreneurship and Individual Responsibility in *Jamie's Ministry of Food. European Journal of Cultural Studies*, 13, pp. 307–22.

Jackson, P., M. Watson, and N. Piper, 2013. Locating Anxiety in the Social: The Cultural Mediation of Food Fears. *European Journal of Cultural Studies*, 16(1), pp. 24–42.

Larner, W., 2000. Neo-liberalism: Policy, Ideology, Governmentality. *Studies in Political Economy*, 63, pp. 5–26.

Lewis, T., 2007. 'He Needs to Face His Fears with These Five Queers!' Queer Eye for the Straight Guy, Makeover TV, and the Lifestyle Expert. *Television & New Media*, 8(4), 285–311.

Lewis, T., 2008. *Smart Living: Lifestyle Media and Popular Expertise*. New York: Peter Lang.

Lupton, D., 1996. *Food, the Body and the Self*. London: Sage.

McMurria, J., 2008. Desperate Citizens and Good Samaritans: Neoliberalism and Makeover Reality TV. *Television and New Media*, 9, pp. 305–32.

Mennell, S., 1996. *All Manners of Food: Eating and Taste in England and France from the Middle Ages to the Present* (2nd edition). Chicago: University of Illinois Press.

Miller, D., 1998. *A Theory of Shopping*. Cambridge: Polity Press.

Moseley, R., 2000. Makeover Takeover on British Television. *Screen*, 41(3), pp. 299–314.

Moseley, R., 2001. Real Lads Do Cook… But Some Things Are Still Hard to Talk About: The Gendering of 8–9. *European Journal of Cultural Studies*, 4(1), pp. 32–9.

Oliver, J., 2008. *Jamie's Ministry of Food*. London: Michael Joseph.

Ouellette, L. and J. Hay, 2008. Makeover Television, Governmentality and the Good Citizen. *Continuum*, 24, pp. 471–84.

Perrie, R., 2006. Sinner Ladies. *The Sun*, [online] 16 September. Available at: http://www.thesun.co.uk/sol/homepage/news/63611/Sinner-ladies-sell-kids-junk-food.html [Accessed 5 November 2013].

Piper, N., 2013. Audiencing Jamie Oliver: Embarrassment, Voyeurism and Reflexive Positioning. *Geoforum*, 45, pp. 346–55.

Readers' Letters, 2006. *Yorkshire Post*, 20 September, page unknown.

Rich, E., 2011. 'I See Her Being Obesed!': Public Pedagogy, Reality Media and the Obesity Crisis. *Health*, 15, p. 3–21.

Rousseau, S., 2012. *Food Media: Celebrity Chefs and the Politics of Everyday Interference*. Oxford: Berg.

Sender, K. and M. Sullivan, 2008. Epidemics of Will, Failures of Self-esteem: Responding to Fat Bodies in *The Biggest Loser* and *What Not to Wear*. *Continuum*, 22, pp. 573–84.

Shilling, J., 2005. Can Our Hero Slay the Beast of the Turkey Twizzler? *Times*, 25 March, Section 2, p. 2.

Silva, E., 2007. Gender, Class, Emotional Capital and Consumption in Family Life. In E. Casey and L. Martens, eds. *Gender and Consumption*. Aldershot: Ashgate, pp. 141–59.

Sims, P., 2006. Mothers Defy Jamie to Deliver Junk Food Over the Playground Fence. *Daily Mail*, 16 September 16, p. 29.

Skeggs, B., 2005. The Making of Class and Gender through Visualizing Moral Subject Formation. *Sociology*, 39, pp. 965–82.

Slocum, S., J. Shannon, K. Cadieux, and M. Beckman, 2011. 'Properly, with Love, from Scratch': Jamie Oliver's Food Revolution. *Radical History Review*, 110, pp. 178–91.

Stokes, P., 2006. Mrs Chips Takes Orders for the School Dinners Run. Mothers Defy Health Campaign with Fast Food. *Daily Telegraph*, 16 September, p. 5.

Strange, N., 1998. Perform, Educate, Entertain: Ingredients of the Cookery Programme Genre. In C. Geraghty and D. Lusted, eds. *The Television Studies Book*. London: Arnold, pp. 301–12.

Tronto, J.C., 1989. Women and Caring: What Can Feminists Learn about Morality from Caring? In A.M. Jaggar and S. Bordo, eds. *Gender/Body/Knowledge*. New Brunswick: Rutgers University Press, pp. 172–97.

Tyler, I., 2008. Chav Mum, Chav Scum: Class Disgust in Contemporary Britain. *Feminist Media Studies*, 8, pp. 17–34.

Tyler, I., 2011. Pramface Girls: The Class Politics of Maternal TV. In H. Wood and B. Skeggs, eds. *Reality Television and Class*. London: BFI, pp. 210–24.

Tyler, I. and B. Bennett, 2010. 'Celebrity Chav': Fame, Femininity and Social Class. *European Journal of Cultural Studies*, 13(3), 375–393.

Warde, A., 1997. *Consumption, Food and Taste: Culinary Antinomies and Commodity Culture*. London: Sage.

Warin, M., 2011. Foucault's Progeny: Jamie Oliver and the Art of Governing Obesity. *Social Theory and Health*, 9, pp. 24–40.

White, S., 2006. Cruel Dinners: Protesting Mums Supply Meals-on-wheels Junk Food for Children. *Mirror*, 16 September, p. 4.

6 Manning the table

Masculinity and weight loss in U.S. commercials

Fabio Parasecoli

'I have won like a man as long as I have been a man. Now I am losing like one' (Barkley, 2014). These words are part of the message, featuring former basketball player Charles Barkley, that welcomed visitors to the home page of the Weight Watchers' online advertising campaign for men. Weight Watchers, based in the United States, markets products and services to help individuals lose weight and keep it off. In recent years, the international company has attempted to expand its client base to men, shifting from its original in-person model to include a more far-reaching – and discrete – online presence. In doing so, Weight Watchers has successfully created a virtual space where men can deal with their weight issues away from the scrutiny and judgment of others. In this chapter, I will examine the campaign featuring Barkley as an example of how U.S. mainstream media and marketing portrays men's interactions around food and eating. The goal of this analysis is to investigate how representations and discourses about food-related practices and male body image embrace and naturalize a set of expectations, values, and behaviors as constitutive elements of acceptable masculinity.

This study is based on theory and research in men's and gender studies suggesting that masculinities are not monolithic and unchangeable, indelibly etched on men's bodies, but rather socially and historically constructed. Traits that are often understood as inherent and essential to all men are actually in constant transformation, allowing for the emergence of different models indicating that we should speak of masculinities in the plural, also within the same culture (Coad, 2008; Craik, 2009, pp. 116–17). These plural masculinities are the ever-changing outcome of ongoing power negotiations within different discourses, practices, representations, and institutional arrangements (Baker, 2006; Brod and Kaufman, 1994; Flood, Kegan Gardiner, Pease, and Pringle, 2007; Kimmel, Hearn, and Connell, 2005; Whitehead, 2006). Media and communication constitute one of the most important fields where these negotiations take place in post-industrial societies.

The Weight Watchers' media and advertising blitz featuring Barkley places issues of male identity at the center of the communication strategy, a tactic which had been present but not emphasized in previous weight loss campaigns targeting men. Starting from the very name of the campaign, 'Lose Like A Man,' the connection between masculinity, food, and body image sustains each commercial. The choice of Charles Barkley, a popular icon with a professional athlete background (an occupation Americans consider appropriate to men) and a very tall and imposing figure, is meant to erase any fear and embarrassment that men may experience in dealing with weight loss, a preoccupation that literature has shown as traditionally attributed to women (Blood, 2004; Bordo, 1993; Grogan, 2008). Furthermore, although Barkley's race is never directly addressed in the commercials, in American mainstream entertainment black athletes and musicians often present a hyper-masculine veneer that is the result of both complex historical dynamics and subtle yet intentional media representations (Parasecoli, 2008, pp. 116–19).

What do real men eat?

The 'Lose Like A Man' online campaign highlights Barkley's manliness as a sport star and as a champion, while depicting weight loss as winning, resulting from an activity as strenuous and tough as physical exercise. References to confident masculinity are redundant, almost to assuage men's anxiety about the whole process. For instance, the twitter account connected to the campaign, @LoseLikeAMan, allowed clients to 'get tips, observations and quotes from real men on the plan, the Weight Watchers' Men's editorial staff and, of course, Charles.' The word *man* recurs several times in the campaign home page, emphasizing even the gender of the editorial staff dedicated to the website. This may be a necessary disclaimer lest male readers fear that women surreptitiously give them advice about issues they know nothing about and try to impose their priorities and points of view on them.

Immediately after listing Barkley's top five favorite workouts (basketball, golf, mixed martial arts, crossfit, and the elliptical trainer), the campaign home page showcases his food preferences, indicated as his 'Favorite Man Foods':

> There are some foods that men just love to eat, and I was surprised I didn't have to give up any of them in order to lose weight. Here are my 5 favorite man-meals, and ideas for making healthier versions when you're cooking at home.
>
> (Barkley, 2014)

The way the topics are laid out on the webpage suggests a connection between physical activities (ranging from rough-and-tumble to moderate) and food consumption aimed at weight loss. The foods Barkley claims to prefer are yogurt, Brussels sprouts, beans, cauliflower, and eggs, with recipes provided for each. While beans and eggs are part of the normal diets of many Americans, yogurt and vegetables hardly register among the ingredients that U.S. popular culture and media frequently refer to as 'men's food,' such as meat, burgers, chips, pasta, pizza, and such. These foods are presented as 'naturally' craved by men, reaffirming their manhood by the very act of consumption. They are at the same time the building substance, and a performance of masculinity.

An analysis of the discursive mechanisms connecting men and food in the U.S. media goes beyond the scope of this chapter, but even the most cursive and non-scientific foray reveals the role that cooking and eating play in performances of masculinity. For instance, writer Terry Durack (who assures readers that 'No women were approached for information, advice or recipes in the making of this article') claimed in a story about Father's Day foods on goodfood.com:

> The real difference between men and women is what they eat. Men are from Meatland, women are from Planet Chocolate. According to David Katz, a nutrition authority from America's Yale University, gender preferential eating habits have been driven by evolution. He claims that men need more protein than women to build muscle mass and, as hunters, view meat as a reward.
>
> (Durack, 2013)

The website of *Southern Living* magazine included smoked and barbequed meats, fried foods, and chili among the 'Dad-Approved Foods' (Dad-Approved Foods, 2014). *Taste of Home* listed men's food according to the occasions when they may be consumed, such as hunting, fishing, tailgating, camping, and cooking with beer, all of them usually considered as typical leisure activities (Guy Food, 2014). Meat, especially grilled meat, is frequently perceived as the quintessential male food in mainstream American culture, recurrently connected with the outdoors, special events, and conviviality. Barbequing is acceptable for men because it is clearly separated from the domestic sphere of the daily care-giving and nurturing, often marked as feminine (Bentley, 2005; Roth, 2005). As sociologist Jeffery Sobal observed:

> singular models of masculinity gender-type foods as masculine and feminine, suggesting that men and women "do gender" by consuming gender appropriate foods. Meat, especially red meat, is an archetypical masculine food. Men often emphasize meat, and women often minimize, in displaying gender as individuals.
>
> (Sobal, 2005, p. 135)

For this reason, meat eating has acquired connotations that at times appear at odds with gender equality. In her 1990 publication *The Sexual Politics of Meat: A Feminist-Vegetarian Critical Theory,* feminist activist Carol Adams claimed that the consumption of meat reflects the patriarchal mindset that tends to treat others as objects for male supremacy (Adams, 2010). Some of the most popular diets among American men, including the Paleo and the Atkins diets, focus mainly on protein, referring to science – from evolutionary biology to physiology – to assert the lack of nutritional value of vegetables (Parasecoli, 2008, pp. 85–102; Stibbe, 2004).

Marketing often exploits and exaggerates a supposedly natural connection between men and their appetite for abundant and appropriately masculine foods.

In certain mythologizing approaches within U.S. mainstream cultures, it appears relevant that men consume food that, as the idiom goes, 'sticks to the ribs.' In other words, masculine food is supposed to be to be nutritious and satisfying, making a man strong and keeping him full without much concern about the effect on his looks. In reality, body image plays a crucial role in defining dominant masculinity models. In heterosexual contexts, strong – and even large – male bodies have traditionally been perceived as expressive for sexual potency, independence, and power (Bronski, 1998). However, in recent years also men have showed widespread – and increasingly public – preoccupation with the way they look. The expression 'the Adonis complex' has been coined to indicate the more pathological forms of obsession with excessive weight, fitness, and muscularity (Pope, Phillips, and Olivardia, 2000). These anxieties have become a common feature in masculine practices and identification processes for both heterosexual and homosexual males. Some extreme behaviors are now referred to as 'orthorexia,' a fixation with correct eating defined by the National Eating Disorders Association as an:

> unhealthy obsession with otherwise healthy eating. . . . Orthorexia starts out as an innocent attempt to eat more healthfully, but orthorexics become fixated on food quality and purity. They become consumed with what and how much to eat, and how to deal with 'slip-ups.' An iron-clad will is needed to maintain this rigid eating style. Every day is a chance to eat right, be 'good,' rise above others in dietary prowess, and self-punish if temptation wins (usually through stricter eating, fasts and exercise). Self-esteem becomes wrapped up in the purity of orthorexics' diet and they sometimes feel superior to others, especially in regard to food intake.
>
> (Kratina, n.d.)

Men and weight loss in media

Within this changing cultural and social framework, media focusing on dieting plans for men turn into an arena where different ideas and practices about masculinity, health, and body image are negotiated. Men can pay attention to their caloric intake, but they are not supposed to make a big deal of it. At any rate, they should still be allowed to eat the food that is culturally appropriate for them and that they are supposed to prefer – at least according to marketing and advertising. Consequently, promoting diet products to men can be complicated, as indicated by a very successful 2009 Superbowl commercial for Pepsi Max, a zero calorie cola that tried hard to avoid stigmatization usually attached to diet drinks. We see young men hit by golf clubs, bowling balls, and electric shocks, all inflicted by careless male friends during shared activities varying from sport to partying and do-it-yourself construction, all clearly marked as masculine. The voiceover states: 'Men can take anything, except the taste of diet Cola. Until now. Pepsi Max, the first diet cola for men.' The commercial ends with one of the injured men nursing a bruise with a chilled can of Pepsi Max, saying: 'This is good.'

If marketers have a hard time defining diet products in terms of masculinity, the dynamics are even more complicated with weight loss plans, which require more intentionality and clear goals. In recent years, several campaigns for slimming programs have focused on men, attempting to introduce acceptable masculinities that, although securely positioned in the mainstream, allow some room for diversity. The Subway company promoted the success story of Jared Fogle, an overweight man who came to media attention for having lost weight by eating Subway sandwiches. In 2000 Fogle became the official spokesperson for the franchise, appearing in commercials, visiting Subway stores, and giving talks on healthy eating. The image of his slim self next to his old enormous jeans has become iconic, building on a long-lasting tradition of before-and-after images proving a person's weight loss. Though his presence in Subway's marketing was phased out around 2008, Fogle still had his own page on the franchise website for some time, under the title: 'Jared's Journey: The Journey Continues,' celebrating his accomplishments, including his participation in the 2010 ING New York City Marathon (Subway, n.d.).

In 2010 actor and comedian Jason Alexander, who rose to fame as George Costanza in the sitcom *Seinfeld,* became the first male spokesperson for the weight management company Jenny Craig in the campaign 'Jen Works for Men.' The transition was marked by the joint appearance of the previous female spokesperson, actress Valerie Bertinelli, in the campaign featuring Alexander. In the commercial celebrating his 15-pound loss in 5 weeks, the two are in a kitchen, where Alexander shows Bertinelli how the Jenny Craig consultant taught him how to prepare his own food so he's never hungry. When they sit down to eat, he finds a tight swimming trunk instead of a napkin next to his dish. In a following commercial (20 pounds in 9 weeks), the two discuss how overweight persons tend to wear their shirt untucked to cover their shape (the 'untucked shirt of denial'), while playing pool. After Bertinelli pockets a ball in a hole located right in front of Alexander's crotch (a jibe to his masculinity), she shows him another tight swimming trunk, this time directly suggesting that he show off his new shape. Eventually, in a commercial titled 'Baring It All' (30 pounds in 18 weeks), Alexander launches himself in a Broadway number that ends with him stripping to his underwear and even, as the title suggests, 'baring it all.' The trunk flies into Bertinelli's shocked face, who whispers: 'You're my hero' (Jenny Craig, 2010).

Although the campaign centers on the concept that Jenny Craig is good for men too, the choice of Jason Alexander, an actor with a Broadway background who is not particularly imposing in the first place, may have generated a mixed message. His strip dance, despite the admirations it elicits in Bertinelli, did not fall within the framework of mainstream American masculinity, which generally pushes Broadway tunes and dancing to the women's (and gay men's) sphere of preference. Furthermore, the presence of a female who has already been through the ordeal and thus can play the role of guide could have been perceived as emasculating and embarrassing to men.

These communication strategies and the varying representations of the relationships between men and their food in media and advertising are commercially motivated and definitely do not aim at reframing acceptable gender relations. However, they seem to go in the same direction as theoretical approaches in gender, masculinity, and queer studies that have emphasized the existence of different kinds of masculinities defined not only by gender and sexuality, but also by age, race, ethnicity, and class.

When applied to the examination of practices, discourse, and representations, this framework of analysis can reveal structures of patriarchal domination that express themselves in forms of men-on-men oppression and exploitation. These tensions also reverberate in complex ways in the relationships that men experience with women and in society at large. The increasing visibility of gay men – from drag queens to leather aficionados – in American society, their influence of aspects of material culture such as fitness, fashion, and design, as well as the growing numbers of 'metrosexuals,' heterosexual men interested in clothes and other forms of consumption that were previously the exclusive domain of gay men, suggest that various models of masculinity are emerging in a negotiated plurality that reflects actual and imagined power relations.

These dynamics point to men's studies pioneer R. W. Connell's insight that various models of masculinity can coexist and compete for preeminence, and that specific circumstances determine the emergence of one or more of those models as hegemonic, guaranteeing its dominant position over other men and the subordination of women (Carrigan, Connell, and Lee, 1985; Connell, 1995; Connell, 2000; Connell and Messerschmidt, 2005). As a consequence, contrasting masculinities find themselves in relationships not only of competition, dominance, and subordination (Segal, 1990), but also of complicity with each other – in order to partake of the benefits deriving from patriarchy, befalling on all men – and sometimes of marginalization, as the result of the attempts to deprive certain categories (homosexuals, ethnic minorities, etc.) of these benefits.

In this context, it is important to observe that power and order are not always imposed on the subject from the outside, but are materialized through embodied norms and regulations (Marcel Mauss's 'techniques of the body'), which are in turn the object and the result of negotiations among individuals, communities, and social configurations, fully expressing the tension between structure and agency. To be fully effective, power relationships and the negotiations that continuously modify and define them need to be embodied and performed by the individual subjects. This is also the case for food consumption, which is nevertheless framed simply as a biological necessity. The way we categorize and experience our physical needs, the way we choose, store, prepare, cook, ingest, digest, and excrete food, are far from being neutral or natural. Furthermore, eating patterns change according to age, geographical location, and other social factors, engulfing a physical act in a whole network of power relations and negotiations, which can express themselves through embodied and societal norms.

Male bodies and the way they look reflect these tensions. What, how, and with whom men eat and the results on the way they look and feel about themselves impact their social standing. Masculinity is performed, regulated and reinforced by daily gestures and practices that often risk being overlooked because of their supposed naturalness. These subtle forms of control are usually excluded from the public discourse to create the illusion of the neutrality of the body that, instead, constitutes the battlefield for cultural, social, and political struggles. Moreover, power needs to be reinforced with legitimacy so that individuals and communities voluntarily adhere to its dictates and rules. The narratives, prescriptions, objects, and practices that facilitate the transmission and diffusion of expectation and ideas in the public space of communicative and material exchanges all share an ideological function, in the sense that they tend to reinforce specific values and goals for society at large.

Sir Charles loses like a man

These power negotiations surrounding masculinities are visible in media and advertising representations of men's relationship with their bodies and what they choose to eat. This is also the case for the Weight Watchers' 'Lose Like A Man' campaign, featuring Charles Barkley. Without losing his status as a sports icon, Barkley successfully managed to transition from top basketball athlete to sports commentator and overall media personality, as his hosting of the comedy show Saturday Night Live in 2010 and 2012 suggests. However, after his retirement from the NBA in 2000, the player gained a sizable 100 pounds, exemplifying the plight of many men who used to be active in their youth and then fall into eating habits and routines that damage their health while impacting their fitness. It is arguably for this reason that Weight Watchers approached him for a campaign specifically targeting men, a move that required a drastic change of brand personality for a company that had been mostly geared towards women. Barkley officially started the 'Lose Like A Man' campaign on December 25, 2011, at the end of the holiday season when Americans are most worried about having gained weight because of the festive and abundant meals. He informed the press that he had already lost 14 pounds by himself before going on the program in September, and in the following months he lost 27 more pounds (Charles Barkley Weight Watchers, 2011; Game On, 2011).

In the first commercial released on December 25th, 2011, Barkley dealt with the perception of his persona as a role model (Charles Barkley in Weight Watchers, 2011). The camera alternates between close-ups of the former athlete and full figure shots, which reveal his whole body. He wears a black buttoned-down shirt open at the neck, black pants, and black shoes. His body contour blends discretely into the black background, not letting viewers actually see what his shape is like. The dark clothes against the dark backdrop are discretely expressive of the desire that many overweight men feel to avoid being noticed. Barkley talks directly to the audience, looking straight into the camera, over the soundtrack of electric rock guitar:

> I am still not a role model, but maybe I can change that. Maybe if I tell you I am losing weight and getting healthy, you see that you can too. Maybe if I say I stopped making excuses and started making progress, you'd do the same.

> And maybe, if I told you I was doing it with Weight Watchers, you'd join me. Lose like a man! Go to Weightwatchers.com and join for free.
>
> (Charles Barkley in Weight Watchers, 2011)

Barkley is trying to establish a connection with other men in his same situation, creating a brotherhood of peers. Although he acknowledges his celebrity status, he clarifies that he has become worthy of being a role model only after taking responsibility for his choices. Others can do the same by joining the all-male online community, which allows enough discretion and independence to individuals without depriving them of a support system. Furthermore, men, who are supposed to enjoy gizmos and technology, should appreciate the connection through the Internet. From the beginning, the 'Lose Like A Man' campaign underlines important elements that are quite revealing of the ideology and the practices surrounding food and body image, as well as their connection to masculinity. First of all, the effort to lose weight does not make you any less of a man. The presence of a 6′6″ pro-athlete with a deep, raspy voice and a masculine demeanor is very reassuring. If he has been doing it, any other man could without too great of an impact on his self-esteem. The electric guitar sound also suggests ruggedness, underlining that the campaign is for men. Second, the decision to watch one's food intake is about health, not body image. Men are supposed to focus on the important stuff, without being steered by supposedly feminine traits like vanity. Third, losing weight requires determination and constancy, two attributes that are often identified as appropriate for men in American mainstream culture. Although there is a reference to losing weight together and joining a program, the emphasis on collective effort, one of the main strategies implemented by Weight Watchers since its inception, is not as relevant as in the traditional campaigns aimed at women (Hendley, 2003; Vignali and Henderson, 2008).

Of course, to be effective the campaign had to be present on different kinds of media. The campaign rebounded from the Internet to TV and printed media, using both visual and textual material, in a self-maintaining loop designed to achieve visibility across various platforms. On January 3, 2012, just a few days after the launch, Barkley appeared on the *Daily Show with Jon Stewart,* an extremely popular comedy show, talking candidly about his sartorial problems related to his considerable girth (Comedy Central, 2012).

More videos appeared marking Barkley's loss of 23 pounds (most of the videos are not available online any longer). In the one titled 'Charles Weighs In on Man Food,' filmed in the same style as the first one (although shots from the waist up were introduced), the athlete starts by reciting a list of supposedly 'man foods':

> Pizza, burgers, wings, meatballs, ribs, steak, meatballs. Men don't want to live without them. With Weight Watchers, we don't have to. They've got a plan for men that teaches you how to eat your favorite food and still lose weight. I have lost 23 pounds already and I haven't given up a thing. Don't worry guys, you can lose your weight and keep your meatballs.

The foods named in the commercial are usually not part of daily meals prepared at home. They are rather snacks (pizza, wings) or meat dishes (burgers, ribs, steak) that are often ordered in, consumed at restaurants and sports bars or enjoyed on occasions like outdoor barbeques or tailgating parties. The only outliers are meatballs, which appear twice in the list although they belong to a more homey style of food. Barkley mentions them at the end, as the single item men would be particularly worried about keeping. The sexual innuendo is quite clear: you won't be castrated while following this diet. Yet the underlying theme in much of the communication about weight loss in the United States still makes its appearance: it is possible to slim down without renouncing any favorite foods. This idea, pervasive also in campaigns geared towards women, translates itself into plans providing customers with ready-made meals that include popular items like desserts and snacks. Substitutions of fat-free or sugar-free products for the real ones are acceptable, with the goal of changing one's lifestyle as little as possible, without rethinking one's eating patterns or nutritional needs. The food industry embraces and promotes this approach through communication, marketing, and the constant introduction of new diet products (Gough, 2007).

At the 23-pound weight loss mark, Weight Watchers released two more videos: 'Charles weighs in on nicknames' and 'Charles weighs in on D.I.Y,' all shot in the same style as the previous ones. In the first, we see Charles confess:

> I have been called fat boy. The bread truck. I've been called Sir Conference. And I earned those names. Now I want to talk about a different name: Weight Watchers. That's right, I said it. Weight Watchers has a plan for men that is easier than you think. I've lost 23 pounds and I'm just getting started. If the round mound can slim down, so can you.

Barkley can employ self-deprecating irony because he has assumed responsibility on his choices. At the same time, the commercial implies that overweight men deserve to be made fun of because it is ultimately their fault if they did not take the necessary measures to stay fit. The third commercial at 23 pounds focuses instead on the fact that although self-reliance is a trait that men embrace, when it comes to weight loss, some assistance is necessary:

> Every guy thinks they can install the new dishwasher themselves . . . till they blow up the kitchen. And every guy thinks they can lose weight on their own. But look around, guys: it ain't working. Weight Watchers has a plan for men that helped me lose 23 pounds already and gives me the tools to keep losing. I tried doing it myself, didn't really go too well.

The use of light irony and colloquial speech pattern make the commercial more enjoyable, taking some of the stigma and the heaviness off the topic. At the same time, the athlete underlines the need for men to rely on external help unless they end up doing something silly, like destroying a kitchen while trying to repair a home appliance.

Overall, the first series of commercials builds on widely shared assumptions about masculinity to make dieting not only acceptable for men, but also a way to express their determination and their inner strength even if starting a slimming plan means going beyond pride to get help. The message is effective precisely because it refers to concepts and practices that most audiences are likely to be familiar with – the idea of manly foods and the social behaviors that surround them, the bullying often connected with overweight individuals, individuality and autonomy as male characteristics, the emphasis on results. Although they appear to make fun of these cultural elements, by including them in the communication strategy the commercials acknowledge their power and their role in determining acceptable masculinities.

A second series of commercials were released to mark the 42-pound weight loss. While the shots, the editing, and the soundtrack are the same as in the previous ones, Barkley's body is lighted more clearly, showing its newly gained fitness in smarter clothes. For instance, in the commercial entitled 'Charles weighs in on game day,' instead of a loose buttoned-down black shirt, the former athlete wears a tight-fitting black vest over a long-sleeved exercise shirt. This is not the same man audiences saw a few months earlier: he is more comfortable with his body and it has reclaimed some of his athlete persona. In the spot, Barkley discusses one of the most visible aspects of American mainstream masculinity: the passion for *watching* sports, which is not necessarily paired with sport practice or exercise. However, men are not berated for not performing physical activities – after all, a diet plan already requires effort in itself, adding exercise would be cruel:

> This is a known fact: watching sports is better with food. Basketball, football, even tennis is better with hot wings. And that's why Weight Watchers is perfect for guys. It shows us how to eat at stadiums and sports bars and still lose weight. I'm down 42 pounds and I go to games for a living. Seriously, I know hot dog vendors by name.

It is important that men can continue eating the food they used to, so they can participate in social activities and not be ostracized for passing on man-appropriate dishes and snacks. Dieters can take some jabs from Barkley, but they would not want to make their dieting too visible to their peers outside of the Internet community. In 'Charles weighs in on real meals,' the spokesperson ridicules other diet plans that require individuals to eat special pre-packaged or ready-made meals that can be bought in stores or even delivered to their door:

> No carbs, fat-free shakes, meals in the mail? You can try to lose weight like that, or you can eat like a normal person. With Weight Watchers, I have already lost 42 pounds and still eat what I want. Just smarter. So unless you prefer getting your food from a mailman, I'd say it's game over.

In 'Charles weighs in on knuckleheads,' Barkley goes for the kill, using the 'F' word: fat.

> No guy wants to ask for directions. Not on the highway, and not on how to lose weight. And here we are, fellas, lost and fat. But Weight Watchers has a plan for men that is really easy and really works. I've already lost 42 pounds by just learning to eat smarter. Nobody likes a knucklehead, guys. Especially a fat one.

While the apparent theme is again excessive self-reliance based on the illusion men can do everything without asking for help, the adjective 'fat' appears twice in the commercial. Men are openly called fat and advised to stop being knuckleheads, to the point that a connection between being fat and obtuse is suggested. The approach is much more direct and rougher than in other commercials featuring Barkley. However, the overall themes that this second series of commercials focuses on are not that different from the first series. Once again, the core of the communication is the negotiation between the need to stick to what is perceived to be mainstream and 'normal' masculinity and the desire to lose weight and achieve a fitter body.

In March 29, 2012, Weight Watchers released the most unusual commercial in the 'Lose Like A Man' campaign. The black backdrop and the rock guitar riffs are the same, but this time we see Barkley in a long-haired wig, flashy earrings, jewelry, a décolleté little black dress, and open-toed pumps, walking towards the camera. Full-body shots of him posing with his hands on the hips in the most unfeminine way possible or shaking his hair, alternate with close-ups where he addresses the audience:

> I hear some of you guys out there think that Weight Watchers is just for women, even though I, Sir Charles, have been telling you that Weight Watchers helped me lose 42 pounds and counting, and I can still eat man food, like steak and pizza. So, if this is what I gotta do to get your attention, take a good look. But my eyes are up here, guys.
>
> (Weight Watchers Commercial: Charles Barkley, 2012)

While he says this, he pretends to get the audience's attention from this chest to his face. Mimicking supposedly feminine movements in ways that actually highlight his muscles, powerful structure, and total lack of gracefulness, while joking on the idea that men tend to focus on women's physical attributes rather than on their faces, Barkley manages to make his masculinity even more evident. He is not in drag, trying to appear womanly. Delivering his lines in a deep and raspy voice, he is just a man wearing women's clothes to get a laugh from his buddies.

To underline that Barkley is just a guy like any other and that the weight loss plan does not work only with celebrities, a commercial with a completely different visual and verbal approach was added to the campaign reassuring viewers that anybody can 'lose like a man' (Weight Watchers Online Featuring Charles Barkley, 2012). Barkley is in a bar. He is smartly dressed in a blue suit and vest, dark purple shirt unbuttoned at the neck, with no necktie. He is relaxed, with one arm over the back of a chair, a tall glass of red beer in front of him. We know he's been on a diet, and the presence of the beer glass reinforces the idea that weight loss does not require men to give up what they enjoy and their social activities. 'People always say to me: Chuck, you've lost all this weight on Weight Watchers, but you're rich, famous and handsome (the hint of a satisfied smirk appears on his face). But what if you are an ordinary guy, like these two nobodies.' He nods to the side, and the camera moves to two young men sitting at his table, holding the pictures of their previous overweight selves. Erik 'Nobody' and Eddie 'Nobody,' both handsome and younger than Barkley (possibly in their early 30s) are wearing button-down shirts, their sleeves rolled up, definitely less sharp than Barkley, but still looking well put together. They look at each other and then Eddie says, 'Yeah well, we lost a ton of weight on Weight Watchers and we did it all online.' Erik: 'There is a plan, and it's just for guys' – Barkley looks at them approvingly – 'It has apps to keep you on track and even cheat sheets for beer and steak.' Barkley, looking into the camera: 'See? Even a nobody can do it.' He looks at Erik and Eddie again. 'We have as many rings as you do,' says Eddie, referring to the championship rings given to the members of winning teams. 'That hurt,' answers Barkley, and continues, 'Lose like a man! Go to WeightWatchers.com and join for free,' over images of a computer on a black background next to the words DO IT in all caps and white lettering, then a smartphone next to DO IT ENTIRELY, and finally a tablet next to DO IT ENTIRELY ONLINE, followed by the Weight Watchers name, end date of the special offer, and the website address. Once again, the technological aspects of the campaign and the cool tools offered to men to track their food consumption (and even cheat) are presented as an enticing plus.

This commercial marks a shift in the communication strategy. While Barkley was on the Weight Watchers' website for some time, he is no longer an active spokesman. However, in the public perception, he remains connected with the dieting plan and he is often asked questions about it. As a matter of fact, while discussing the experience on the Conan O'Brien late night show, he jokingly claimed that for every 35 pounds a man loses, he gains an inch in penis size, and in fact he gained two inches (Charles Barkley Tells Conan, 2013). Although Weight Watchers may not have endorsed these observations, it is relevant to notice that the connection between dieting and masculinity is maintained, even enhanced. This element has remained at the core of the following campaigns for male weight loss plans.

Erik and Eddie from the Barkley's commercial also had blogs on the company website, where they share their everyday experiences (Weight Watchers, 2012). Moreover, new commercials have been produced that showcase the success stories of non-celebrities including regular guys Daniel, Pete, and Matt. Daniel, a former military and a techie-geek, points out how proud he was when he could first wear 32″ waist pants, and wonders how his wife manages to keep her hands off his new hot self. Pete says he looks with sadness at pictures of his former overweight self with his son, and admits that he spends quite some time in front of the mirror admiring his abs. Matt appears with his wife Meg, who observes how she actually prepares the meal making sure that her husband stays on track. This commercial blatantly harps on some of the most common stereotypes about mainstream heterosexual couples like the husband who is only able to microwave food and the wife getting her way on daily matters. It is noticeable that the campaign does not feature black men, while at the same time Weight Watchers had black female R&B singer and actress Jennifer Hudson as a spokesperson in its campaign for women. It is unclear whether this choice indicates that black men are not considered as valuable targets for dieting programs or if black men lose their value as spokespersons once the hyper-masculinity connected to specific athletes or performers is out of the picture.

The new Weight Watchers commercials for men show a major shift from Barkley's campaign not only because they focus on everyday people, but because they introduce elements that had been accurately kept out of the previous communication strategies. Although the protagonists of the commercials are undoubtedly manly and lead normal men's lives, or maybe precisely because of that, themes like vanity, sexuality, and couple life are introduced. This time, men have no qualms getting help, even from their apparently overbearing wives. Men care about their looks, the clothes they wear, and their public perceptions. It seems like Weight Watchers is exploring new territory in its communication from men. May this be the result of surveys and analysis of the previous campaign impact? Is the company signaling awareness of changing perceptions and behaviors among American men about their gender identity and performance? In the evolution of Weight Watchers campaigns, masculinities reveal their transitory nature and their continuous transformations, making visible already accepted elements while at the same time introducing and reinforcing new ones.

Conclusion

The inner workings of masculinities are made visible in media and marketing, becoming part of a shared repository constituting the field of social activity that can be defined as collective imagination. The analysis of food- and body image-related media, which provide a supposedly neutral space for representations of masculinity, can identify a set of recurrent traits that outline different but acceptable ways of being men. Thanks to the increasingly fast exchange of information and the global outreach of U.S. media industries, these models are offered to audiences both in the cultural environment in which media are produced, but also in many other places around the world. Eating and dieting offer consumers untapped opportunities to reflect upon the ideas and behaviors that constitute adequate and respectable masculinities. By dealing with an aspect of life – food consumption – that is often perceived as simply innate and motivated by biological needs, media participates in the production and naturalization of cultural bias, social dynamics, and power hierarchies by turning the norms, negotiations, and tensions that underpin them into entertainment and marketing.

References

Adams, C., 2010. *The Sexual Politics of Meat: A Feminist-Vegetarian Critical Theory*. New York: Continuum.

Baker, B., 2006. *Masculinity in Fiction and Film: Representing Men in Popular Genres, 1945–2000*. London and New York: Continuum.

Barkley, C., 2014. *Marketing*. [online] Available at: <http://www.weightwatchers.com/templates/Marketing/Marketing_Utool_1col.aspx? pageid=1316051> [Retrieved 11/21/2014].

Bentley, A., 2005. Men on Atkins: Dieting, Meat and Masculinity. In L. Heldke, K. Mommer, and C. Pineo, eds. *The Atkins Diet and Philosophy*. Chicago: Open Court, pp. 185–196.

Blood, S. K., 2004. *Body Work: The Social Construction of Women's Body Image*. New York: Routledge.

Bordo, S., 1993. *Unbearable Weight*. Berkeley and Los Angeles: University of California Press.

Brod, H., and M., Kaufman, 1994. *Theorizing Masculinities*. Thousands Oak, CA: Sage.

Bronski, M., 1998. The male body in the Western mind. *Harvard Gay and Lesbian Review*, 5(4), pp. 28–32.

Carrigan, T., R.W. Connell, and J. Lee., 1985. Toward a new sociology of masculinity. *Theory and Society* 14(5), pp. 551–604.

Charles Barkley in Weight Watchers: Being a Role Model, 2011. [video online] Available at: <http://www.youtube.com/watch?v=MuE_rtwmxX0> [Retrieved 11/21/2014].

Charles Barkley Tells Conan Penis Grew 2 Inches, 2013. *Huffington Post*. [video online] April 2013. Available at: <http://www.huffingtonpost.com/2013/04/05/charles-barkley-tells-conan-penis-grew-2-inches_n_3020055.html> [Retrieved 11/21/2014].

Charles Barkley Weight Watchers, 2011. *LA Times*. [online] Available at: <http://articles.latimes.com/2011/dec/13/news/la-heb-charles-barkley-weight-watchers-20111213> [Retrieved 11/21/2014].

Coad, D., 2008. *The Metrosexual: Gender, Sexuality, and Sport*. Albany: SUNY Press.

Comedy Central, 2012. *The Daily Show with Jon Stewart*. [video online] January 2012. Available at: <http://www.thedailyshow.com/watch/tue-january-3–2012/charles-barkley> [Retrieved 11/12/14].

Connell, R. W., 1995. *Masculinities*. Berkeley: University of California Press.

Connell, R. W., 2000. *The Men and the Boys*. Berkeley: University of California Press.

Connell, R. W., and Messerschmidt, J., 2005. Hegemonic masculinity: Rethinking the concept. *Gender and Society*, 19(6), pp. 829–859.

Craik, J. 2009. *Fashion: The Key Concepts*. Oxford and New York: Berg.

Dad-Approved Foods, 2014. *Southern Living* [online] Available at: <http://www.southernliving.com/food/holidays-occasions/20-man-food-recipes> [Retrieved 11/21/2014].

Durack, T. 2013. What men like. *GoodFood.com*. [online] Available at: <http://www.goodfood.com.au/good-food/cook/what-men-like-20130819–2s62t.html> [Retrieved 11/21/2014].

Flood, M., J. Kegan Gardiner, B. Pease, and K. Pringle, ed. 2007. *International Encyclopedia of Men and Masculinities*. London and New York: Routledge.

Game On, 2011. *USA Today*. [online] Available at: <http://content.usatoday.com/communities/gameon/post/2011/12/charles-barkley-gets-serious-about-losing-his-round-mound/1#.Up0ADI2QaIQ> [Retrieved 11/21/2104].

Gough, B., 2007. 'Real men don't diet': An analysis of contemporary newspaper representations of men, food and health. *Social Science & Medicine,* 64, pp. 326–337.

Grogan, S., 2008. *Body Image: Understanding Body Dissatisfaction in Men, Women and Children*. New York: Routledge.

Hendley, J., 2003. Weight watchers at forty: A celebration. *Gastronomica,* 3(1), pp. 16–21.

Jenny Craig, 2010. *Baring It All.* [video online] Available at: <https://www.youtube.com/watch?v=UVQzVPbwfkY> [Retrieved 11/21/2014].

Kimmel, M., J. Hearn, and R. W. Connell, 2005. *The Handbook of Studies on Men and Masculinities*. Thousand Oaks, CA: Sage.

Kratina, Karin, n.d. *Orthorexia-Nervosa*. [online] Available at: <http://www.nationaleatingdisorders.org/orthorexia-nervosa> [Retrieved n.d.].

Parasecoli, F., 2008. *Bite Me: Food in Popular Culture*. Oxford and New York: Berg.

Pope, H., K. Phillips, and R. Olivardia, 2000. *The Adonis Complex*. New York: Touchstone Books.

Roth, L., 2005. "Beef. It's what's for dinner": Vegetarians, meat-eaters and the negotiation of familial relationships. *Food, Culture & Society,* 8(2), pp. 181–200.

Segal, L., 1990. *Slow Motion: Changing Masculinities, Changing Men*. New Brunswick: Rutgers University Press.

Sobal, J., 2005. Men, meat, and marriage: Models of masculinity. *Food & Foodways*, 13(1), pp. 135–158.

Stibbe, A., 2004. Health and the social construction of masculinity in Men's Health Magazine. *Men and Masculinities,* 7(1), pp. 31–51.

Subway, n.d. *Jared's Journey*. [online] Available at: <http://www.subway.com/subwayroot/freshbuzz/Jared/default.aspx> [Retrieved 11.21/2014].

Guy Food, 2014. *Taste of Home*. [online] Available at: <http://www.tasteofhome.com/guy-food> [Retrieved 11/21/2014].

Vignali, C., and S. Henderson, 2008. Weight watchers: Social event centered marketing. *Journal of Food Products Marketing*, 14(2), pp. 99–113.

Weight Watchers, 2012. This is how I roll. *Weight Watchers Blog*. [blog] August 27, 2012. Available at: <http://community.weightwatchers.com/Blogs/UserBlog.aspx?blogid=1037608&createyear=2012&createmonth=8> [Retrieved 11/22/2014].

Weight Watchers Commercial: Charles Barkley, 2012. The NBA on ESPN. [online] Available at: <https://www.youtube.com/watch?v=rBo8F9S5EHU> [Retrieved 11/21/2014].

Weight Watchers Online Featuring Charles Barkley, 2012. *Ispot.tv*. [video online] October 2012. Available at: <http://www.ispot.tv/ad/7LC8/weight-watchers-online-featuring-charles-barkley> [Retrieved 11/21/2014].

Whitehead, S., ed., 2006. *Men and Masculinities: Critical Concepts in Sociology*. New York: Routledge.

7 Homosocial heterotopias and masculine escapism in TV-cooking shows

Jonatan Leer

The Danish show *Spise med Price* (Dine with Price) (2008–) is the most popular cooking show in the history of Danish television. It premiered in 2008 and is hosted by the two brothers James (b. 1959) and Adam Price (b. 1967). They are the sons of Birgitte and John Price, who were both actors and celebrities among the Danish public. John Price was also a well-known food writer and author of the cookbook *Spise med Price* (1973, see Nyvang this volume). The brothers used the same title for their show four decades later. In each of the thirty-minute episodes, we follow the two brothers as they meet up and cook in Adam Price's summer-house for the day.[1]

In this chapter, I will argue that *Spise med Price* can be read as belonging to a tendency in contemporary cooking shows in the UK and Denmark from 2006 to at least 2012 in which men are portrayed cooking in women-free zones, far away from an everyday-like context and far away from women. My argument is that, in this tendency, the male hosts are using food and cooking to 'escape' the constant negotiations of the post-traditional life and to create culinary counter-spaces. In these spaces they bond with other men and play with forms of traditional masculinity in ways that might not be appropriate in the mainstream spaces of modern societies. At the same time, the heterotopic homosociality of *Spise med Price* and the other shows is highly ambivalent, as it is performed in intertwinement with irony and nostalgia.

My understanding of food and cooking as escapism is inspired by the readings of food fairs as sites of carnival proposed by Naccarato and LeBesco (2012, pp. 85–94). There, food practices at food fairs are described as escapism, marking a temporary suspension of the imperatives and norms of everyday life and everyday food practices. Masculine escapism is a specific kind of escapism, in which the absence of women is constructed as a central part of the out-of-the-everyday experience. In this respect, culinary male escapism echoes the traditional gendering of cooking, in which female cooking is constructed as an everyday activity and a care-for-others project, whereas men's cooking is constructed as an extraordinary activity, a care-for-self project (Devault, 1991; Lupton, 1996; Strange, 1999).

This analysis is theoretically inspired by and constructed around Foucault's text 'Of Other Spaces'; and more specifically, by Foucault's key concept, *heterotopia,* 'a kind of actually materialized counter-space,' as well as the less developed concept *heterochronie,* the temporal discontinuity of the heterotopia. The analysis will be in two parts, one on the spatial and one on the temporal dimension of these culinary counter-spaces.

It will be concluded that the analysis underlines that food television is one of the social spaces in which masculine identity is currently being negotiated. This is particularly interesting as masculine identity is at present surrounded by uncertainty. Food television offers a series of very different ways of using cooking to perform both new and old forms of masculinity.

Masculinity in food television

The literature on masculinity in cooking shows seems to suggest that there have been two main models in the Western tradition for portraying men as 'cooking subjects' in cookbooks and in cooking shows (Strange, 1999; Nyvang, 2013). One was the professional chef, who can be dated back to the birth of the cookbook. This figure is associated with the professionalism and pride of this masculine tradition and its great masters. The other model is the hedonistic *bon vivant* figure, celebrating a masculine tradition of cooking as a care-for-self project (Lupton, 1996, p. 136). On television, a flamboyant example of this tradition is Keith Floyd, who 'famously took the cameras out of the kitchen studio and into whichever exotic location he (iconic wineglass in hand) happened to be cooking in' (Rousseau, 2012, p. xvi). Both the chef and the *bon vivant* could be seen as antithetical to the dominant female model in the cooking domain: the housewife, a tradition that, in the UK, was established on television with Marguerite Patten in the postwar era (Moseley, 2009) and upheld by a series of female food personalities thereafter, notably Delia Smith (Strange, 1999). The male models emphasise extraordinary food and a high level of culinary skills, whereas the women are associated with the traditional values of home cooking, family and everyday-oriented food, i.e. the care-for-others role (DeVault, 1991).

These male models were challenged in the innovative show *The Naked Chef,* with Jamie Oliver, in 1999. Jamie Oliver incorporated home cooking into a wider display of his lifestyle in the series: a lifestyle that was urban, dynamic and young. The show demonstrated how home cooking – a traditionally feminine practice – could be a way for a 'modern' man to win success in various social contexts (Hollows, 2003, p. 230).

The Naked Chef has attracted wide academic interest. Several scholars have noticed the show's reframing of the relationship between cooking and masculinity (Moseley, 2001; Hollows, 2003; de Sollier, 2005; Feasey, 2008, pp. 129–135; Milestone and Meyer, 2012, pp. 142–143), and between masculinity and the home (Attwood, 2005; Brownlie and Hewer, 2007). The figure of Oliver in *The Naked Chef* has been associated with the 'new lad' (Jackson, Stevenson and Brooks, 2001), a cultural figure of masculinity appearing in the UK in the 1990s and constructed around 'a space of fun, consumption and sexual freedom for men, unfettered by adult male responsibilities' (Gill, 2003, p. 47). The 'new lad' has often been interpreted as a response to the 1980s 'new man,' a softer expression of masculinity that opened up the masculine role to feminine spaces and practices (Milestone and Meyer, 2012, pp. 116–117). The 'new lad' is echoed in Oliver's figure by his Oasis look and his informal, streetwise mode of expression, but also in his lack of seriousness regarding both himself and his cooking; life and food 'gotta be a laugh!'

However, in other episodes Oliver seems to tone down his 'laddism' and instead embraces 'the new man'; for instance, in the episode 'Babysitting,' where Oliver is looking after his Uncle Allen's kids. In the course of the afternoon, Oliver introduces his nieces to the art of making ravioli. In this situation Jamie comes across as a much more progressive and modern man who is open to, and intelligently combines, care-taking tasks and domestic family cooking. So rather than being one or the other, the show can be understood as 'a negotiation between these competing discourses of masculinity' (Moseley in Brunsdon et al., 2001, p. 39). Indeed, Oliver's position could be interpreted as an example of 'bricolage masculinity' – 'the result of "channel-hopping" across versions of "the masculine"' (Beynon, 2002, p. 6). The show suggests that this kind of balancing act is expected of a man in the post-traditional era, where identity and gender roles are fluid and balanced through 'a day-to-day social behavior' (Giddens, 1991, p. 71). Furthermore, the show illustrates and idealises how cooking can be a way of 'doing' and negotiating a successful post-traditional masculinity (Søndergaard, 1996).

The masculinity-as-escapism tendency proposes a quite different version of culinary masculinity than *The Naked Chef*. Rather than returning to the old models (although similarities with the *bon vivant* can be found), it tries to create masculine spaces around cooking that offer modern men a pause from the complexity of modern life. So whereas Jamie in *The Naked Chef* uses cooking to bricolage or zap across versions of the masculine, adapting his gender performance to a series of contexts and different social spaces in post-traditional life, the men in the escapism tendency use cooking to slip away into spaces where they are no longer bound by the norms and gender-equal ideals of post-traditional life. Here Foucault's concept of heterotopia becomes a helpful tool to theorise and analyse the tendency.

Heterotopia and homosociality in food culture

In 'Of Other Spaces,' Foucault defines the heterotopia in opposition to the utopia as:

> something like counter-sites, a kind of effectively enacted utopia in which the real sites, all the other real sites that can be found within the culture, are simultaneously represented, contested, and inverted. Places of this kind are outside of all places, even though it may be possible to indicate their location in reality.
>
> (Foucault, 1984, pp. 1574–1575)

As suggested by Johnson (2013), the fragmentary and inconsistent character of Foucault's text has certainly led to its being interpreted and applied very freely. Despite these reservations, Johnson acknowledges the usefulness of the concept for the cultural analysis of social spaces with a counter-dimension. As mentioned by Naccarato and LeBesco, food escapism takes place in counter-hegemonic spaces that contrast with the food spaces of 'normal lives' (2012, p. 88), and in my opinion the heterotopia is best understood as exactly that: a space that offers and presents itself as an escape from the constraints and practices of a so-called or imagined everyday life.

In my examination of this 'masculine cooking as escapism' tendency, I find the concept of heterotopia to be particularly useful to describe and understand the gendered spaces of these homosocial escapades. Also, the focus on space and counter-spaces can help in understanding the differences between masculinity and cooking in *The Naked Chef* and *Spise med Price*. As much of the academic research on *The Naked Chef* has emphasised, the novelty of the series was related to the recasting of the domestic kitchen as a male space (Moseley, 2001; Hollows, 2003; Attwood, 2005; Brownlie and Hewer, 2007). Bearing this perspective in mind, it makes sense to take space as a point of departure for the study of masculinity and cooking in food television.

In this chapter I explore a specific kind of heterotopia that I call homosocial heterotopias. Homosociality is traditionally understood as social bonding between persons of the same sex, subtexted by a rejection of homosexuality (Sedgwick, 1985, p. 1). As my focus is on men, this means women-free spaces that make room for male bonding.[2] Homosocial relations have been considered an important part of the construction of traditional masculinity (Kimmel, 1996, pp. 7–8). Furthermore, several studies have shown how sexism and homophobia are crucial elements to homosociality (Flood, 2008; Ihamäki, 2013).[3] In my readings, I have chosen to focus on the ambivalences of such post-traditional homosocial spaces in order to make sense of the complexities of this kind of heterotopia.[4]

In my understanding of heterotopia, I would like to emphasise two elements that are central to the case of *Spise med Price*. First of all, the importance of the feeling of a different era in the culinary homosocial heterotopias, a feeling that is visible through the importance attached to nostalgia and outdated masculine discourses. Second, the heterotopias are spaces where childish imagination can be unfolded. This point is particularly developed in a radio lecture given by Foucault before the publication of 'Of Other Spaces.' Here Foucault mentions that heterotopias can also be 'Thursday afternoons on their parents' bed,' when children, by using their imagination, 'can discover the ocean' or 'bounce on the springs.'[5] The parental bed becomes an emplacement for 'inventing dream-like spaces that are firmly connected to and mirror the outside world; a fearful, playful experiment with the boundaries of space' (Johnson, 2013, pp. 798–799).

This quote draws an interesting parallel with the practice of the culinary homosociality. The children playing alone in the parents' bedroom behave in a way they wouldn't if the parents were present. Hence, the absence of the parents and the parental order is essential to this kind of heterotopia. In the same way, certain ways of doing masculinity and doing food are made legitimate for men by the absence of women in the culinary homosocial heterotopias: these are ways of doing masculinity that would not be tolerable in everyday bi-gender contexts. By analogy with the parents in Foucault's example of the children and the bed, in the construction of these shows women are associated with a certain 'feminine order' – echoing the above-mentioned myth of feminine cooking (e.g. Devault, 1991; Lupton, 1996; Strange, 1999) – that is associated with moderation, self-control, following the norms, family life and behaving correctly. This order does not apply to the homosocial heterotopias.

The Price brothers and the success of *Spise med Price*

Like their parents, the two brothers hosting *Spise med Price* have worked in show business, although, until the launch of *Spise med Price,* they had not been the object of as much public attention as their parents: James Price is a musician and a band leader at the famous Danish summer cabaret *Cirkusrevyen,* while Adam Price is an autodidact writer with several television series on his résumé, most notably the recent international hit *Borgen*. Like their father, the brothers have worked as food writers and restaurant critics on the side, but *Spise med Price* was their first attempt at a cooking show. In each of the thirty-minute episodes, we follow the two brothers as they meet up and cook for the day in Adam Price's summerhouse.

Spise med Price was originally produced for the small and somewhat elitist channel DR2, part of the Danish Broadcasting Corporation. This public-service institution has a long tradition of food programming since the 1960s with the creation of *TV-Køkkenet* (The TV Kitchen). In this show the decor was the same, but the TV chef changed from season to season, sometimes even from programme to programme, resulting in almost 200 cooks assuming the role of studio host between 1966 and 1991 (Boesen, 2000). From the early 2000s, DR2 successfully created a series of low-tech lifestyle/food programmes formed around the lifestyle of characteristic – often somewhat politically engaged – and provocative personalities (Carlsen and Frandsen, 2005). This type of food programming differs from the classic, objective tradition of public-service-oriented food programmes as in *TV-Køkkenet,* but also from more mainstream and commercially oriented programmes such as *Masterchef* and *Who's For Dinner?*. *Spise med Price* can be seen as an example of this DR2 tradition, although it has focused more on the family story and experimenting with the cooking-show genre – perhaps one of the reasons for its success. After two seasons at DR2, the programme moved to the main channel, DR1, with a huge increase in viewing numbers, and has since become one of the most popular cooking shows in the history of DR.[6] The two brothers, previously marginal figures of the Danish celebrity scene, have gained a popularity equal to their parents. The programme is one of the few examples of an original concept with marginal celebrities conquering the primetime slot on a main channel.

Homosocial heterotopias in *Spise med Price*

The father's kitchen

In the first episode, Adam Price compares cooking to urban legends. Each cook builds his recipes and food stories based on the versions of previous generations, and perhaps even other cultures, while each cook adds their own personal touch to the story.[7]

The central legend in *Spise med Price* is the family chronicle. In the introduction to each show, the family tree is exposed through a collage of annotated photos on a refrigerator door. This introduction narrates how their mother was a terrible cook who 'only cooked in film' when one of her roles required it. By contrast with this untraditional mother, the father Price passed his 'intense love for vigorous cooking with cream and butter' (Price and Price, 1973/2012, p. 3) on to his sons. They still consider his cookbook 'a kind of Bible.' The brothers' show, *Spise med Price,* is in many ways a homage paid to their father and to the culinary tradition of male food writers of the 1960s of which he was a part. This tradition valorises French country cooking and writes in an animated, hedonistic style, with frequent quotes from the 'grand' French gourmets (see Nyvang's chapter in this book).

This paternal bond that drives the brothers' approach to cooking is framed as distinctively masculine, and we get the impression that the women in the family are not a part of this cooking tradition. The mother couldn't cook for her life; and in an episode on retro food,[8] we are told that though her own mother cooked a wonderful Sunday lunch for the whole family every Sunday for most of her life, on her deathbed she had disclosed that she had always hated to cook!

So not only recipes, but also the intense love of cooking, are passed from father to son(s). It is important to note that this culinary legacy is not only about the sensual pleasures of eating, but a love of cooking as a masculine practice.

*Isolation and anti-*30 Minute Meals

In each episode, the two brothers meet up in Adam Price's summerhouse. This cottage appears to be one big kitchen, filled with gadgets, kitchen machines and stuffed refrigerators. The brothers spend a whole day just cooking together. Often they adventure extremely complicated dishes. For instance, in one episode the brothers cook a *cassoulet,* the famous and very nutritious bean stew from Carcassonne with several kinds of meat and *confit de canard.*[9] This dish is truly a project that demands quite a bit of preparation: sophisticated shopping, for several untraditional cuts of pork and cured specialties such as *ventrèche,* a Pyrenean type of bacon; the preparation of the confited duck in advance, which in itself takes several days; and the fully invested brothers even make their own home-made sausages, which requires the acquisition of a sausage stuffer and a smoking oven. With these preparations accomplished, the actual cooking of the stew can begin! This too is a process which demands an entire day and a multitude of pots and pans for frying the various types of meats and the bean stew sauce, before finally assembling the dish and giving it a couple of hours in the oven.

As the two brothers enjoy the finished meal at the end of the show, Adam Price praises the dish and notes that 'You can really taste all the processes,' and concludes, 'It has been worth all the hard work,' after which James Price corrects his brother: 'But it's not hard work, to me it was great fun.' Cooking is established as a male leisure practice, as opposed to feminine cooking (Lupton, 1996) as exemplified by the cooking-hating granny.

In *The Naked Chef* there is also an aspect of cooking as a fun and care-for-self activity (Hollows, 2003), but here cooking also serves a practical purpose: namely, to increase Oliver's social standing.[10] This aspect is completely absent in *Spise med Price:* the brothers cook in an isolated summerhouse outside of urban sociality, and no guests are expected. It is just the two of them and an enormous pot of cassoulet.

This kind of cooking, where huge amounts of time, know-how and money are invested in a single dish, without any consideration for practicalities, seems to be a kind of anti-*30 Minute Meals* cooking. In the *30 Minute Meals* concept, time-saving and domestic efficiency are core values (Nathanson, 2009).[11] In *Spise med Price* it is quite the contrary: the time invested adds to the value of the dish. Hereby it is also evident that the Price brothers and the *30 Minute* tradition are focusing on two different contexts: the latter offers solutions to busy everyday life in modern families, whereas the Price brothers are focusing on cooking as a fun hobby. *The Naked Chef* could be read as a kind of middle ground, with space for both pleasure and practicality.

The fact that the show is set in a summerhouse clearly relays the wish to dissociate cooking from a practical, everyday-oriented context. The summerhouse kitchen becomes a counter-space, a heterotopia, unlike both *The Naked Chef* and everyday cooking in the family kitchen, where a parent has to stress through the supermarket on the way home from work and quickly cook a decent and healthy meal before the kids' bedtime. In the heterotopic summerhouse, such considerations are absent from cooking, as well as all the discourses that organise such considerations.

A space for anti-healthy cooking

In the summerhouse, cooking is also dissociated from the normativity of food discourses and from what the food sociologist Claude Fischler has called 'health régimes.' The health régime 'designates a set of rules to be followed . . . [and] exercises a control over everyday living, the body and its behavior' (Fischler, 1988, p. 290). In contemporary societies, Fischler argues, the fat body has lost its potential positive connotations (wealth, abundance) and the slim body has become an important status marker signifying control and self-discipline (Fischler, 1989).

In the summerhouse, the food régimes and the ideals of slim food are continually and very explicitly transgressed, as the two brothers continue to uphold their father's cooking style with lots of cream and butter. In the episode 'Grease' (season 1, episode 4), the brothers only eat deep-fried dishes, where fat is closely connected to profound pleasure. One example is when the brothers eat James Price's favourite dish, *scampi fritti* (deep-fried battered shrimp), this time replaced by the exclusive Norwegian lobster, with a mayonnaise-based tartare sauce as the only condiment. James Price tells us how he ate this dish every day on his first trip to Italy as a kid with his parents, and Adam Price remarks, half moaning, 'Insane luxury to eat langoustine like this!' The excessiveness of this dish both in regard to expense and healthy eating recommendations is essential to this moment of homosocial happiness. Furthermore, the fat dish is closely connected to memories of childhood, and the fat is attributed a nostalgic quality as well. However, the show also underlines that the brothers seem to accept that this style of cooking can only be realised in this culinary counter-space.

A competition space

A recurrent theme in the show is the brothers' competition over which one of them has the largest and most expensive kitchen machines and gadgets: the so-called arms race. To take the lead in this race, one of the brothers has to surprise the other with a new machine, for instance when James Price arrives with his own vacuum-packed, smoked sausages for the cassoulet. To make these he has brought not only a smoking oven, but also a vacuum-packing machine that he ironically remarks is a natural part of any decent household. Adam has to make a counter-attack, and in a subsequent episode he surprises James as he brings his own hand-built Portuguese garden-oven, which can generate a heat of 350 degrees Celsius and is ideal for making pizzas.[12] This game continues from episode to episode. In the concluding episode of season two, the director makes a weapon-count of their individual collections.

It has been argued that competition is an important part of much homosociality and that it serves to establish – and to challenge the established – hierarchies, but can also work as 'a source to solidarity, comradeship and mutual affection' (Meuser, 2007, p. 38). Although the competitive element is toned down by heavy use of irony, it is still a central part of the Price brothers' homosociality, and something that marks this space out as different from an everyday context, where such childish games would be gauche. It is only with a brother that you can do something like this, and hereby the gastronomic arms race is perhaps above all – to use Meuser's words – a sign of their conjunction and their mutual affection, reflecting their shared love for food and the bond they form, through cooking, with each other and with their father.

All the heterotopias mentioned by Foucault also are places in which specific bonds are generated between the people sharing the counter-spaces. These individuals are, by their entrance and presence, members of a specific club, even partners in crime. The proximity between the brothers is a core element in *Spise med Price.* Cooking here could even be read as a way of performing proximity, as a ritual affirming the bond of brotherhood.

A fantasy space

In episode five of *Spise med Price,* the two brothers cook a classic Italian dinner.[13] As James Price prepares the dessert, *tiramisu,* he is looking for the Amaretto, an Italian almond liqueur, to moisten the ladies' fingers that constitute the base of the dessert. He asks his brother: 'Is it at the big band's studio?' This is a rather strange question, considering that the two brothers are in a small summerhouse. But James's reply is affirmative, and Adam walks across the kitchen and exits through a door. In the next clip, Adam Price is in a big music studio where Danish Radio's big band is recording the music for the show. In the trumpet section there are some bottles of booze. Following a small dispute, Adam is permitted to borrow the half-empty bottle of Amaretto. He leaves the room the same way he came in, and returns to the kitchen complaining about the big band's massive alcohol consumption.

This magic-door gimmick is a recurrent feature on the show, and in season two the door becomes a way of entering the entire universe of the Danish Broadcasting Corporation: the brothers drink wine with a famous news host, they borrow milk from the leading cultural programme, *Smagsdommerne,* and they offer a chilli taste sample to the weather presenter who, due to the dish's spiciness, is unable to speak during a broadcast.

These comic stunts destabilise the realistic feeling of space that has been a key element of lifestyle television. In this genre, space is often idealised or romanticised, but always only just enough to enchant the viewer without making the space seem unreachable or surreal. The magic door is a fundamental transgression of this ideal, adding a fantastic element to the space of the summerhouse. This kitchen – like the parents' bed for Foucault's child at play – is a magical space where everyday objects can turn into magical things by the use of imagination.

A fragile space

In several episodes, the culinary homosocial heterotopia is exposed as a fragile space that depends on both brothers' equal investment in creating the right atmosphere. In this sense, the heterotopic composition is constantly in danger of falling apart. For instance, in an episode on pork,[14] it is made clear that the two brothers share a common love for this animal; however, James Price's passion differs from Adam's in some aspects. Being allergic to fur, James Price has always wanted a baby pig as a pet, given that pigs don't have fur. During the programme he disappears mysteriously several times, and we later discover that he has been visiting a nearby field where organic pigs and their piglets are enjoying life. Here he spends time with a new-found love: a cute baby piglet.

Adam Price is clearly annoyed by his brother's devotion to this tiny animal and warns him several times: 'Don't make friends with it, then you won't eat it!' Finally, Adam convinces his brother to come back to the house and have a pig party, as they are going to taste all the fatty pork dishes they have been cooking throughout the entire day. They have been especially worked up about a gigantic *porchetta* that has been barbecued slowly for more than four hours on an enormous gas grill. As they are about to eat this roast, the main attraction of the day, Adam seizes the moment to get back at his brother. As James is about to dig into a *porchetta* sandwich, Adam reminds him that this is 'the very soul of the pig.' James is visibly disturbed by this sentence, and after a few bites he wants to go back to his piglet, to feed it some leftover fruit.

The moment the entire episode has been building up to – the consumption of the long-awaited *porchetta* – becomes a kind of anticlimax and a staging of the fragility of the culinary homosocial space, exposed by the baby piglet.

Homosocial heterochronies

In this second part of the analysis, I will move the focus from the counter-spatial to the counter-temporal dimension. My overall point in this part of the analysis will be that the homosocial heterotopia evokes, and is constructed through, a series of outdated discursive repertoires of masculinity. At the same time, these discourses are recycled, as we shall see, with great ambivalence and irony.

'Grandmother's Egg' and family nostalgia

The very first dish the Price brothers cook in the very first episode is called 'Grandmother's Egg.' As the brothers prepare the dish they retell the anecdote that the family has attached to this dish. John Price – their father – as a child was poor and fatherless. As he and his mother had little to eat, they would go to bed early on cold winter evenings and read cookbooks to satisfy their hunger and dream of all the nice things they would eat once they were rich.

One of the dishes she made for him – when it was especially cold or when he was ill – was 'Grandmother's Egg': a boiled egg with butter croutons. Grandmother's croutons, roasted in butter, were the most luxurious thing they could imagine, and the more butter, the more luxurious the dish. To make this simple dish smarter and more extravagant – like the food they were reading about in the cookbooks – it would be served in a wineglass. The Price brothers also point out that this dish ideally should be eaten by someone who feels a bit unwell and in need of attention. The person eating the dish should be lying down, and the cook should serve the dish saying: 'Ohh, poor you.'

The brothers tell the story with great pathos, and the dish is presented as the ultimate comfort dish; the viewer can easily picture the scene, the mother comforting her son with this treat. However, there is also a touch of irony: the brothers emphasise the comic aspect of the grandmother's petit-bourgeois ideal of food, underlined by the curious idea of serving a boiled egg in a wineglass. By their ironic pathos, the brothers also expose to the viewer that, rather than a true story, this anecdote is a discourse recounted to them by the flamboyant actor-father about his childhood and his relationship with his mother.

Cooking, in the heterotopia, becomes a way of revitalising and portraying family members who are no longer alive, as well as the affective relationships between them. This, however, is not done through glorification of the family members, but by commemorating them through an ironic performance in which food and story-telling are interwoven.

The way in which the brothers relate to their parents and grandparents might best be described by the term 'reflexive nostalgia' (Boym, 2001). In her book, *The Future of Nostalgia,* Svetlana Boym proposes a distinction between restorative and reflective nostalgia. Restorative nostalgia regrets modernity and longs for the restoration of 'the original stasis' (p. 49), a mythologised version of the past. Reflexive nostalgia has a more dreamy fascination with 'the individual and cultural memory,' but does not see restoration of the past as possible or desirable. Hence reflexive nostalgia can have a more ironic and ambivalent approach to the past, but its nostalgic narrative is not one shared chronicle of home(land), but rather 'ironic, inconclusive and fragmentary' (p. 50). It takes the past less seriously and less literally, but the past is still a referent for affection.

Lots of butter: Reflective nostalgic revival of the **bon vivant!**

This reflective, nostalgic approach towards the past can be found not only in the context of the family chronicle, but also of a series of cultural figures and tropes of the past, most notably the figure of the male *bon vivant.* Around the 1960s, the Danish foodscape was invaded by a series of male chefs, who, inspired by the French gourmets of the nineteenth century, promoted a leisure-oriented cooking style and recast the kitchen as a masculine hobby room. This school was a reaction to the housewife-isation of the food media. The housewife had been a central figure of Danish food media since the beginning of the nineteenth century and continued to dominate the first cooking shows on Danish TV in the 1960s. The housewife figure represented an everyday-oriented, practical, economically responsible ideal of cooking and eating (see Nyvang's chapter in this book). In opposition to this, the gourmet cooks prepared complicated and luxurious food and established cooking as a masculine practice and domain. John Price was one of these gourmands; another was the restaurant reviewer Conrad Bjerre-Christensen, who, accompanied by his quiet sidekick Aksel Larsen, was the popular host for *TV-Køkkenet* in the 1960s and 1970s.

In each episode of *Spise med Price,* several short clips are incorporated of their father, John Price, and the show with Conrad and Aksel, as the duo was baptised by the public. One clip is especially important as it is shown in every episode of *Spise med Price* – every time the brothers use butter. It is a one-second clip showing Conrad putting butter in a pan as he prescribes: 'Lots of butter!' With these sequences edited into the Price brothers' cooking demonstrations, the viewer is constantly reminded of the importance of the legacy of the tradition of male chefs, and their cooking style with plenty of cream and butter.

Furthermore, the clip alludes to a significant conflict concerning Conrad and Aksel's appearances on TV. The couple infringed the public-service institution's ethics by appearing, in their *TV-Køkkenet* costumes, in a series of commercials for

Danish butter, with the slogan: 'Use butter, that's what we do!' The campaign was created by an advertising agency that was part of the Danish agricultural society. They were very pleased by the use of butter in the TV kitchen – especially as butter on the Danish market was increasingly threatened by margarine. Danish Radio Broadcasting, however, was not pleased at the use of a public-service institution as a marketing platform, particularly as it concerned unhealthy products. Aksel and Conrad were fired.

It is notable that butter became the centre of this conflict, as butter was already a distinctive sign of the gourmand tradition as an opposition to the housewife tradition, in which margarine was used happily. DR took a stand in the fight between the gourmands and the housewife tradition by firing the duo – a stand against the extravagant, self-indulgent gourmand.

The 'lots of butter' phrase has become synonymous with the *Spice med Price* show among the Danish public. By this means the brothers Price reactivated the old gendered distinction between hedonistic masculine cooking versus practical and healthy feminine cooking, positioning themselves in the gourmand tradition. In the recycling of masculine hedonism, they seem to have placed themselves in opposition to the housewife tradition and also to its modern versions in the form of a series of female cooking-show hosts such as the "fat exorcist" Anne Larsen, who was widely discussed by the Danish public around the time *Spise med Price* premiered (Christensen, 2007). However, it is also clear that the show revived the masculine tradition in a homosocial counter-space. Put in Foucauldian terms, the brothers acknowledge that this is a heterochronic discourse, and that it can only be unfolded in a heterotopic context.

Cooking and killing

The homosocial space is, however, not just about nostalgia. The summerhouse also appears to be a site for evoking discursive repertoires and practices that are inappropriate in ordinary social situations. The most obvious example is the explicit fascination with violence in relation to cooking. Adam Price tells the story of how his father let him play with a live lobster as a child. He named it Helene. After an hour, the father took it from him and said to the young boy: 'Now Helene will be baptised.' Helene was dumped into a pot filled with boiling water. 'Then I learned that cooking is killing,' Adam ends the story. The connection between cooking and killing is a recurrent theme, often expressed through the use of black humour; notably the episode on lamb, 'Silence of the Lambs' (season 1, episode 3), just like the book and movie about the sophisticated cannibal Hannibal Lecter and the serial killer Buffalo Bill, who skins his female victims.

Like these serial killers, the brothers are fascinated by cutting up meat and playing with it: outstandingly, when they perform a ballet with two deboned chickens in a marionette theatre, to a soundtrack of kitsch polka music.[15] In some instances the meat is also sexualised, for example when the brothers prepare the *porchetta* in the episode on pork: they start with a deboned piglet that is massaged with herbs and firmly laced up to a rolled joint, before Adam impales it on a spear. As he penetrates the piece of meat with his spear, he emphasises: 'The tighter the better!' and James comments, with a corny smile, 'This is not a programme for children.' Then Adam proudly lifts the penetrated animal with one hand at each end of the spear and starts using it as a barbell, while the theme from *Rocky* plays.

As these examples make clear, the summerhouse is a place where it is permitted to play with the violent dimension of cooking. The Price brothers seem to take pleasure in breaking this taboo, and they could be interpreted as a part of a new 'carnivorism' in popular gastronomy (Parry, 2010). This tendency operates with a gendered distinction between the feminised, ridiculed farmed animals (pp. 383–384) and the heroic, masculine act of slaughtering (pp. 385–387). Consistent with this tendency, the brothers' meat handling activates an archaic mythology of masculinity as dominance; however, as they do so in this case with an ironic twist, it renders the sexism ambiguous.

Other examples of masculine escapism in cooking shows

As already mentioned, there are additional examples in contemporary food television of cooking like that of *Spise med Price* becoming a means of masculine escapism. Another example of this trend is the Danish show *Nak og Æd* (Whack and Eat, also DR2 2010–), in which the hunter Jørgen Skouboe and the cook Nikolaj Kirk escape into the wilderness for a couple of days to kill an animal and cook it over an open fire. They play with the primordial hunter-gatherer masculinity in complete isolation from women. The aesthetics of the show clearly emphasise this action-masculinity by ironically reproducing the aesthetics and motifs of the action-series genre by overlaying graphic texts such as 'Camp 0731' or 'Pick-up Point 1548.'

Even one of Jamie Oliver's more recent shows could be read as a variant on this tendency, namely *Jamie at Home* (Channel 4, 2006–2008). In this show, home is no longer the metropolitan bachelor apartment, but a big house deep in the Essex countryside. Oliver is living in this pastoral heaven with his wife Jools and their three children. Although the series is set at home, the wife and kids are completely absent. The show focuses on Oliver and his garden, and on how he grows and cooks his own vegetables. Oliver's only company in the show is the tranquil gardener Bryan, who explains to Oliver and the viewer the little tricks behind growing your own vegetables. In exchange, Oliver treats Bryan to his cooking, and the series exposes many warm moments between the two men. The garden is enclosed by a great wall, distinctly encapsulating this little homosocial microcosm, liberated from the burdens of work and family life.

Oliver's garden materialises as a counter-site to the urban, social arena so fundamental to *The Naked Chef*. The Jamie Oliver of *Jamie at Home* has escaped the social scene that the young Oliver tried to conquer through cooking. As an opposition to the utopian bricolage masculinity of *The Naked Chef*, *Jamie at Home* seems to propose an escape into a heterotopic homosociality. The new show also clearly reiterates – as did *Spise med Price* and *Nak og Æd* – a series of heterochronic discourses of pastoral idyll and a return to a non-abstract, seasonal and, literally, more rooted way of eating and being. Although *Jamie at Home* doesn't have the same ironic attitude towards the heterotopic construction as *Spise med Price* and *Nak og Æd,* it idealises a homosocial space where men are taking some time of their own, with cooking.

Conclusion

This chapter has argued that *Spise med Price* is an example of what I have referred to as a masculine-cooking-as-escapism tendency, and that *Nak og Æd* and *Jamie at Home* can also be considered examples of this tendency. What these shows have in common is that they portray men cooking and bonding in a space isolated from an everyday-like context. The homosocial spaces are constructed around imagination, fun and nostalgia, and they appear to be explicitly impenetrable to women, or to dissolve themselves if penetrated by women or disturbed by feminised objects (such as the baby piglet in *Spise med Price*). The shows appear to operate with an underlining dichotomy between the feminine and the masculine. The feminine is constituted by all that is absent in the programme and relates to the complexity of modern life: urbanity, constant negotiations of gender identity, seriousness and health and food normativity; whereas the masculine is associated with all that is present in the homosocial heterotopias: isolation, nature, simple living, fun and a lack of seriousness.

Both *The Naked Chef* and the programmes in the masculine-cooking-as-escapism tendency can be read as ways of using cooking to respond to the uncertainty of contemporary masculinity. In *The Naked Chef,* Jamie used cooking as a way to navigate between different repertoires of masculinity and create a mobile masculinity that lived up to the imperative of flexibility and negotiability of post-traditional masculinity. *Jamie at Home, Spise med Price* and *Nak og Æd* do not show how cooking can be a way of navigating in a complex, late-modern sociality. Rather, the shows portray spaces in which men take a break from the complexity of the bi-gender and gender-equal society. In these programmes, cooking becomes – as Foucault also remarked was essential for the heterotopia – a site for compensation.

The analysis also points to the fact that food television is one of the social spaces in which the masculine identity currently is being negotiated. This is particularly interesting as the masculine identity at present is surrounded by uncertainty, and food practices are presented, in these shows, as 'tools' to help men navigate in this uncertain age. Food television does not, however, offer a one-size-fits-all solution, but a series of competing discourses on ways of using cooking to perform new and – maybe most importantly – old forms of masculinity in hegemonic and heterotopian spaces. Despite the apparent plurality, it seems that gendered distinction between masculine and feminine cooking (and eating) has not disappeared. These distinctions are perhaps less explicit and less one-dimensional than in the 'old days' – with gourmand-chef versus housewife – but they still guide the identity bricolages of contemporary food practices.

Notes

1 This chapter only deals with the first three seasons, which all take place in the summerhouse. The subsequent seasons are a bit different also regarding the use of space, as the show uses a greater variety of locations.
2 Homosociality has in the literature predominantly been understood as men's bonding with other men, but could just as well concern women's bonding with other women, e.g. Binhammer (2006).
3 The key point in Sedgwick (1985) is to challenge this idea through readings of English novels from 1750–1850 in order to 'hypothesize the potential unbrokenness of a continuum between homosocial and homosexual' (Sedgwick, 1985, p. 1).
4 This reading strategy is inspired by the approach to the 'new lad' magazines presented in Jackson, Stevenson, and Brooks (2001).
5 Quoted in Johnson (2013) (p. 798). This radio segment is available in French on http://foucault.info/ (last visited 10 April 2014).
6 The Danish newspaper *Politiken* (18 April 2010) confirmed that *Spise med Price* had the highest number of viewers for a cooking show on Danish television (a little over one million). See http://politiken.dk/mad/madnyt/ECE949172/broedrene-price-erobrer – danmark/ (last visited 20 Nov. 2014).
7 *Hanen og Ægget* (The Rooster and the Egg), season 1, episode 1.
8 *Kære Lille Mormor* (Dear Little Granny), season 2, episode 8.
9 *I Tykt og Tyndt* (Through Thick and Thin), season 1, episode 9.
10 As noted by Naccarato and LeBesco (2012), much food-related television can be understood as offering food-related solutions to potential social problems and 'a variety of ways of being in relation to food that might be thought to increase one's social standing' (Naccarato and LeBesco, 2012, p. 41).
11 This concept was launched in 2001 by American chef Rachael Ray on the Food Network (Nathanson, 2009).
12 *Sulten i Provence* (Hungry in Provence), season 2, episode 2.
13 *Italiensk for Let Øvede* (Italian Level 2), season 1, episode 5.
14 *Grise med Price* (Pig 'n' Price), season 3, episode 3.
15 *Tivertifald Kylling* (Chicken, of Course), season 2, episode 6.

References

Attwood, F., 2005. Inside Out: Men on the Home Front. *Journal of Consumer Culture*, 5 (1), pp. 87–107.

Beynon, J., 2002. *Masculinities and Culture*. Buckingham: Open University Press.

Binhammer, K., 2006. Female Homosociality and the Exchange of Men. *Women's Studies*, 35 (3), pp. 221–240.

Boesen, U., 2000. *Tv-køkkenet – De bedste opskrifter og historier i mere end 25 år*. København: DR Multimedie.

Boym, S., 2001. *The Future of Nostalgia*. New York: Basic Books.

Brownlie, D., & Hewer P., 2007. Prime Beef Cuts: Culinary Images for Thinking Men. *Consumption, Markets & Culture*, 10 (3), pp. 229–250.

Brunsdon, C., Johnson, C., Moseley, R., & Wheatley, H., 2001. Factual Entertainment on British Television: The Midlands TV Research Group's '8–9 Project'. *European Journal of Cultural Studies*, 4 (1), pp. 29–62.

Carlsen, J., & Frandsen, K., 2005. *Nytte-og livsstilsprogrammer på dansk tv*. Århus: Arbejdspapir 133 fra Centre for Kulturforskning, Århus Universitet.

Christensen, D. R., 2007. Fedtuddrivelse som mytologiseringspraksis. En kulturanalyse. *DIN. Norsk Tidsskrift for Religion og Kultur*, 1 (1), pp. 89–100.

De Sollier, I., 2005. TV Dinners: Culinary Television, Education and Distinction. *Continuum*, 19 (4), pp. 465–481.

DeVault, M., 1991. *Feeding the Family: The Social Organization of Caring as Gendered Work*. Chicago: University of Chicago Press.

Feasey, R., 2008. *Masculinity and Popular Television*. Edinburgh: Edinburgh University Press.

Fischler, C., 1988. Food, Self and Identity. *Social Science Information*, 27 (2), pp. 275–292.

Fischler, C., 1989. *L'homnivore*. Paris: Odile Jacob.

Flood, M., 2008. Men, Sex, and Homosociality: How Bonds between Men Shape Their Sexual Relations with Women. *Men and Masculinities*, 10 (3), pp. 339–359.

Foucault, M., 1984. Des Espaces Autres. In M. Foucault (2001), *Dits Et Ecrits II (1976–1988)*. Paris: Gallimard, pp. 1571–1581.

Giddens, A., 1991. *Modernity and Self-Identity: Self and Society in the Late Modern Age*. Stanford: Stanford University Press.

Gill, R., 2003. Power and the Production of Subjects: A Geneology of the New Man and the New Lad. In B. Benwell (Ed.), *Masculinity and Men's Lifestyle Magazines*. Oxford: Blackwell Publishing, pp. 34–57.

Hollows, J., 2003. Oliver's Twist: Leisure, Labour and Domestic Masculinity in the Naked Chef. *International Journal of Cultural Studies,* 6 (2), pp. 229–248.

Ihamäki, E., 2013. Homosociality in the Sexscapes of Sortavala, Russia. *Norma: International Journal of Masculinity Studies*, 8 (1), pp. 27–41.

Jackson, P., Stevenson, N., & Brooks, K., 2001. *Making Sense of Men's Magazines*. Cambridge: Polity Press.

Johnson, P., 2013. The Geographies of Heterotopia. *Geography Compass*, 7 (11), pp. 790–803.

Kimmel, M., 1996. *Manhood in America*. Oxford: Oxford University Press.

Lupton, D., 1996. *Food, the Body and the Self*. London: Sage Publications.

Meuser, M., 2007. Serious Games: Competition and the Homosocial Construction of Masculinity. *Norma*, 2 (1), pp. 39–51.

Milestone, K., & Meyer, A., 2012. *Gender and Popular Culture*. Cambridge: Polity Press.

Moseley, R., 2001. Real Lads Do Cook … But Some Things are Still Hard to Talk About: The Gendering 8–9. *European Journal of Cultural Studies*, 4 (1), pp. 32–39.

Moseley, R., 2009. Marguerite Patten, Television Cookery and Postwar British Femininity. In: S. Gillis & J. Hollows (Eds.), *Feminism, Domesticity and Popular Culture*. London: Routledge, pp. 17–31.

Naccarato, P., & LeBesco, K., 2012. *Culinary Capital*. New York: Berg.

Nathanson, E., 2009. As Easy as Pie Cooking Shows, Domestic Efficiency, and Postfeminist Temporality. *Television & New Media*, 10 (4), pp. 311–330.

Nyvang, C., 2013. *Danske kogebøger: 1900–1970. Fire kostmologier*. Copenhagen: University of Copenhagen.

Parry, J., 2010. Gender and Slaughter in Popular Gastronomy. *Feminism & Psychology*, 20 (3), pp. 381–396.

Price, J., & Price, A., 2012. Forord. In J. Price (1973), *Spise med Price*. København: Lindhardt og Ringhof, pp. 3–4.

Rousseau, S., 2012. *Food Media*. New York: Berg.

Sedgwick, E. K., 1985. *Between Men: English Literature and Male Homosocial Desire*. New York: Columbia University Press.

Søndergaard, D. M., 1996. *Tegnet på kroppen*. København: Museum Tusculanum Press.

Strange, N., 1999. Perform, Educate, Entertain: Ingredients of the Cookery Programme Genre. In: C. Geraghty & D. Lusted (Eds.), *The Television Studies Book*. London: Arnold.

Part II

Practices of food and media

8 'I (never) just google'

Food and media practices

Karen Klitgaard Povlsen

Niels is 44 and single. He lives in a small flat in Copenhagen, with simple furniture but expensive computer equipment. He had forgotten our interview appointment, and was still at work when I rang his doorbell. When he turned up an hour later, he made an excuse – he had marked our appointment in several electronic calendars, even on his iGoogle start screen, but had forgotten it. Some years ago, Niels suffered severe brain damage due to meningitis. Niels presents himself as atypical in all matters, as he has hardly any short-term memory left. He uses his computers to support his memory at work and at home – and also in relation to food. Niels subscribes to email newsletters from the big supermarket chains, and he uses these to remember food ingredients and dishes:

> Well, yes, then I get a bit inspired, read – oh God yes, tenderloin exists, what was it one could do with that? Then I google and find a recipe for tenderloin steaks. I use this and then the daily feeds about food and recipes, which food is good for your brain, for instance, the brain likes nuts and almonds and chocolate.
>
> (NS: 187, 197)[1]

Niels seems an extreme case in several respects. Researchers argue, however, that an extreme case makes evident findings that are present in more typical cases in normalised or naturalised forms that make them more or less invisible until highlighted by the extreme case (Flyvbjerg, 2006; Neergaard, 2007; Yin, 2009). This is why Niels is chosen to introduce this chapter: his case suggests a possible intersectional pattern of media-use in everyday life related to food, consisting of health/disease, education, age and gender. Niels's case is one of five similar but in some respects rather diverse cases in the empirical sample that forms the basis of this chapter. They form a flexible complex of demographic factors such as age, gender and class. The complex can only be understood as whole, not in its singular parts; and, unlike traditional consumer segmentation *à la* Bourdieu, it cannot be generalised to other fields of consumption, such as fashion or cars (Bourdieu, 1984; Warde, 2005).

The question posed in this chapter is thus which patterns can be identified in everyday cross-media-use related to food – *beyond* individual user profiles.

Foucault's concepts of heterotopia and heterotopology (Foucault, 1997) seem to offer a framing that is fluffy enough to make room for the necessary complexity; but they may also offer a preliminary understanding of some of the different forms of complexity or differing user patterns. For this purpose, my analysis is also inspired by insights and methods from practice theory (Warde, 2005) and from the anthropology of media practices (Couldry, 2004, 2010; Bräuchler et al., 2010). I shall return to these theoretical inspirations in detail below. Rather than taking a media-centric approach, I present the topics of media and food and the interrelations between both fields of practice. I close with a presentation of my empirical cases, as an example of how cross-media-use in relation to food may be understood using Foucault's terms, heterotopia, heterotopology and heterochronism. But first I shall introduce the important discussion on mediatisation in current media research.

Mediatisation and everyday practices

The patterns that emerge from the interview cases are, as already suggested, not easily recognisable as collective profiles or segments. They also resist traditional audience segmentation in relation to mass media such as television (Katz and Liebes, 1984; Povlsen, 1999; Schroeder et al., 2003), or lifestyle segmentation (Bourdieu, 1984). A similar problem was discussed by Bird more than ten years ago as a problem of proving media effects:

> We cannot really isolate the role of media in culture because the media are firmly anchored into the web of culture, although articulated by individuals in different ways. . . . The 'audience' is everywhere and nowhere.
>
> (Bird, 2003, pp. 2–3)

Today, with the interactive possibilities offered by digital media, audiences have become users and co-producers of media. This even accentuates Bird's point that individuals do not use or consume media in any 'predictable, uniform way' (Bird, 2003, p. 3). Couldry argues that a practice perspective would liberate media research from the problems of how to prove media effects and allow researchers to address non-media-centric research questions:

> we should open our lens even wider to take in the whole range of practices in which media consumption and media-related talk is embedded, including practices of avoiding or selecting out media inputs, . . . as practices oriented to media they are hardly trivial.
>
> (Couldry, 2010, p. 40)

However, this has only recently begun to impact on media research, where the debate has been concentrated on theories of mediatisation (Krotz, 2007; Schulz, 2004) versus 'mediation' to understand what was going on in increasingly 'media-saturated societies' (Ang, 1996). While mediation has long been used as a term for media communication or representation, mediatisation on the other hand is a term for a focus on longer historical transformations of the institutions of culture and society (Hjarvard, 2009; Livingstone, 2009; Couldry, 2012; Livingstone and Lunt, 2014; Hepp et al., 2015). To prove and understand such transformations, however, media researchers need to understand everyday lives, at home or in institutions and media-producing companies. This need has led to a new interest in anthropological methods and practice theory in media research. In the field of audience studies, Couldry, among others, has stressed the importance of anthropological methods for media studies, arguing that the long personal interview and observations of everyday habits allow us to look at media-use as one ordinary practice among others (Couldry 2004, 2010; Couldry et al., 2007). This was the methodology used to produce the empirical data presented in this chapter.

We have never before had so many media, so easy to access. Hartmann talks of media-rich environments at home, at work and in urban areas (Hartmann, 2011). Media and media-use are domesticated to such an extent that we do not really notice our media-uses, because they are a seamlessly embedded and routinised practice among other practices such as eating. Media devices have become normal objects, normal furniture, at home (Silverstone and Hirsch, 1992; Couldry, 2004, 2008; Hartmann, 2006, 2010; Hollows, 2008). We use different devices when we listen to music with a TV on but without sound while we game, or seek information on a tablet or smartphone. We often use not one media, but two, three or more – we are cross-media-users. Complex media-use patterns are therefore structured by, and in turn structure, our daily lives. This makes the influence of single media hard to pin down except in general terms of how media have become an integral part of people's lives (Couldry, 2009; Thompson, 2013). Which is why we need to study the daily media practices.

Practice theory focuses on saying as doing, in the sense that discourses or sayings are not only themselves doings that construct certain habitual life-worlds (Bourdieu, 1984), but also doings in the sense that the sayings perform social roles and relations (Goffman, 1957; Butler, 1990). These performances themselves are socially regulated through discourses in media and in the bio-politics of late modernity (Foucault, 1978).[2] Reckwitz defines practice as:

> a routinised type of behaviour which consists of several elements, interconnected to one another: forms of bodily activities, forms of mental activities, 'things' and their use, a background knowledge in the form of understanding, know-how, states of emotion and motivational knowledge.
>
> (Reckwitz, 2002, p. 249)

Important here is the perception of media-use as a part of enacted practices in an everyday context (Couldry, 2004, 2010; Warde, 2005). To frame these practices, and to find patterns in them, I looked to Foucault's writings on heterotopia.

Heterotopia and heterotopology

In 1967 Foucault wrote his short text 'Of Other Spaces: Utopias and Heterotopias' (Foucault, 1997) on how to understand societies by explanations of space and place. He defined space as extensions and networks of transitions, parallel to Castells' later understanding of the information age of the digital media revolutions as a network society with spaces of flow (Castells 1996).

Foucault did not mention media in his text, however, but focused on the concept of utopia – literally, 'no place.' He understood utopia as an unreal place that is outside all places, while heterotopias – literally 'otherness places' – are actually localisable, his examples being asylums or churchyards. Foucault uses the term even in relation to discursive spaces (Johnson, 2013, p. 791), but in this context I concentrate on his example of the mirror as a mixed experience of utopia and heterotopia, because the parallel between mirror and media is obvious:

> Between these two, I would then set that sort of mixed experience, which partakes of the qualities of both types of location, the mirror. It is after all, a utopia, in that it is a place without a place. . . . At the same time, we are dealing with a heterotopia. The mirror really exists and has a kind of comeback effect on the place that I occupy: starting from it, in fact, I find myself absent from the place where I am, in that I see myself in there.
>
> (Foucault, 1997, p. 333)

The gaze into the mirror, into the virtual space on the other side of the mirror, may be paralleled to the gaze into the media when we use them. Just as with the mirror, one can withdraw oneself from the media and reconstitute oneself in reality. Just as with the mirror, media-use has the potential to become a heterotopia – at the same time real and virtual, because media offer the possibility of being absent from the place where we actually are.

Important here is the notion of media as a double articulation of material and immaterial objects. The use of media does not constitute specific locations, although all media also exist physically in space. Media-use does, however, constitute a net of heterotopias that come into existence when media are used by individuals. These heterotopias are often described in the media as various kinds of content, such as food, taste and lifestyle, which also constitute the media spaces for heterotopology as places for descriptions of heterotopias. Digital media are also heterochronisms, because the internet is an enormous archive (Finnemann, 2005). In this context all three 'functions' are relevant, but I shall focus on media-use as emergent heterotopias, or in Peter Johnson's words:

> heterotopian sites do not sit in isolation as reservoirs of freedom, emancipation or resistance; they coexist, combine and connect. They are not stable entities: they are contingent qualities. . . . 'Heterotopia' is more of an idea about space than any actual place.
>
> (Johnson, 2013, p. 800)

So does the topic of food in the media establish potential heterotopias and 'ideas about space,' as used by the interviewees?

Food in the media

As several chapters in this anthology demonstrate, food practices are currently represented and performed in the media as never before. Food seems omnipresent, not only in real life but also in the media (Miller, 2007; Johnston and Baumann, 2010; Naccarato and LeBesco, 2012). Food studies is an extremely wide field, and the present chapter does not attempt to comprehensively cover the research done here. I shall just mention a few important titles. The media studies field has also expanded, but there has been limited interest so far in the contemporary media research field of food as content in media and food-related media-use. Anthropology, by contrast, has a strong tradition of food studies (among others Levi-Strauss, 1966; Appadurai, 1988; Counihan, 1999; Counihan and van Esterik, 2008). In sociology food is also a distinct topic (i.e. Douglas, 1972; Murcott, 1983; Lupton, 1996; Beardsworth and Keil, 1997; for an overview, see DeSoucey, 2013). Sociologists, historians and anthropologists have often based their work on food as a content in media as if it these mediations were a mirror of reality (Goody, 1977; Mennell, 1985; Warde, 1997). But media representations are subject to construction, and may be interpreted and negotiated in different ways by media-users. This is why qualitative interviews are important for understanding the ways in which contents are appropriated by users.

Cultural food studies, by contrast, have acknowledged media texts at a representational level. TV chefs and magazines have been a focus for cultural studies (Ashley et al., 2004; Bell and Hollows, 2005; Hollows, 2008; Hollows and Jones, 2010a, 2010b; Bell and Hollows, 2011; Leer, 2014). Povlsen has analysed recipes and photographs of food in Danish and German women's magazines in comparison with the fashion pages, inspired by Roland Barthes' semiotic readings of food advertisements (1986); Frances Bonner (2005, 2009) has argued that food has been available on many platforms for many years. Audience or user studies, however, are in short supply: in her qualitative study of Danish women's uses of magazines, Bente Halkier (2010) focuses on the consumption of food more than media. This chapter is therefore one of the first attempts to study how individuals actually use mediated food contents and texts, seeking to identify which media are important for them in relation to food and how media are combined in daily cross-media-use patterns.

The sociologists Johnston and Baumann (2010, pp. 127–171) in their analysis of taste distinctions in gourmet magazines (a discourse analysis in combination with 30 interviews with American 'foodies') demonstrated the importance of media-use for constructing taste distinctions, but they did not focus on how, when or why media are important in the formation of taste distinctions. Nevertheless, they argued that 'foodies' have become 'omnivores' (Peterson and Kern, 1996), not only in the pessimistic paradox labelled by Claude Fischler (1988), but in American foodies' wide range of eating practices and preferences, from ethnic to organic and high gourmet. They presented this as a process of negotiating taste authenticity between hegemonic forms of good taste, e.g. high gourmet restaurants that demonstrate culinary capital (Naccarato and LeBesco, 2012) and more heteronomic taste preferences, e.g. junk food. Naccarato and LeBesco (2012) focus on TV cooking shows and their promises of possibilities of transformation at a representational level. They also discuss online restaurant reviews as a 'broader reframing of the concepts of "authenticity" and "taste" ' (p. 82), but culinary capital also works well in digital media (Naccarato and LeBesco, 2012, pp. 82–83). The question is thus whether the Danish empirical data on cross-media-uses of food contents show us foodies with cultural capital, or whether the heterotopias produced are of quite another kind.

Empirical challenges and methods

The focus on cross-media-use and food preferences in Denmark after 2010 acknowledges that it is hardly possible to find causal relations between two areas of life that are so all-encompassing. The attempt here is therefore to conduct neither a media-centric nor a food-centric study, but rather to focus on how food practices and media practices intersect, and how they are actually enacted and talked about in an everyday context.

The empirical data were produced in two collaborative, externally funded Danish research projects.[3] Both projects were mixed-methods studies, combining a survey with qualitative interviews. In the digital media project *Changing Borders* (2008–11; 1710 answers), ten questions on food preferences and media-food routines were among the 100 questions of the survey. In the organic trust and food project, *Multi-Trust,* a few questions on media preferences were posed among 88 questions concerning organic labels and food consumption (2013–14; 3436 answers). In the *Changing Borders* project, we combined the survey (2009, N: 1710) with 13 individual interviews and one focus group with three participants. Respondents were recruited from the survey *(http://changingborders.au.dk/wp-content/uploads/2012/05/The-Media-Menus-of-Danish-Internet-Users-2009.pdf)* via email addresses. The *Multi-Trust* project consists of the survey, four individual interviews and one focus group of five participants. Again, the respondents were recruited from the survey on trust in organics (2013; N: 5467).[4] As the data here are from 2014, they will be used to put the data from 2010–11 in perspective.

Both surveys showed that the internet, television, and print magazines and weeklies (Finneman et al., 2012; Thorsø et al., forthcoming 2016; Rittenhofer and Povlsen, 2015) were important in everyday media-food use. Differences in use followed familiar patterns of age and gender, with women being slightly more active in relation to food, the elderly watching more TV and the young spending longer time online. Our reason for selecting for individual interviews from the surveys was partly to balance age and gender differences, but also to look at rural/urban differences. Even though the respondents had noted their email addresses, it was in fact difficult to make appointments with them. In both cases we ended up recruiting people who were highly motivated and interested in food or media, but not representative of people in general. Two groups in particular were eager to participate: people with chronic diseases, and nurses. Of the 25 persons interviewed, six had chronic diseases, of whom five were unable to keep up a normal job. Five were nurses – all female. A female musician spent much of her time in the alternative health scene. Five or six participants were very interested in cooking, but, as we shall see, this group was not easy to demarcate. Two were biking and running addicts, and six were very price conscious about food and spent much time surfing the internet and reading advertisements in print media and on TV to find the cheapest prices. This group hardly bought organic food. In the following discussion I shall not comment further on this latter group, because it is well known from consumer studies. All the interviews were transcribed, condensed and coded.

We ended up with 11 men and 14 women aged in their twenties (4), thirties (6), forties (8), fifties (4), and sixties (3). They were educated a little higher than the general population, but only four had a university degree. Several were nurses and teachers, one had no professional training and a few more had limited professional training. The respondents (except for the 'Price-runners') will presented in more detail below.

In the three (all in all, ten persons) qualitative focus-group interviews and 15 individual interviews, we asked about media as well as food preferences and routines. We consequently asked people to tell us what they normally do ('What did you do yesterday . . . ?') in regard to food preferences and media-use, and if possible we also asked them to show us their normal media procedures, for instance on their laptops and smartphones. We visited them at home or at their workplace, and we asked them to show us how and where they normally use their devices or other media, just as we small-talked about food preferences and daily routines.

Although the focus of the two projects differed, both shared an interest in cross-media-use in relation to food, as well as a mixed-method design combining quantitative data (how many, how much) with qualitative findings on why people do as they say they do in daily food and media practices. In both projects, I conducted the interviewing, assisted by PhD students, and participated in constructing the surveys and the results *(www.orgprints.org/multi-Trust)*.

Heterotopia I: chronic disease

From the *Changing Borders* interviews:

Niels S., 44, divorced with two sons, in a qualified flex-job because of brain damage, studied political science and journalism but did not complete the degree.

Birger T., 58, divorced with grown-up children, engineer with an independent business.

Gitte K., 63, married with grown-up children, had a brief professional training. In early retirement because of health problems.

Torben S., 53, lives with his partner, has a university degree and a job in a cultural centre and has an independent travel agency.

From the *Multi-Trust* interviews:

> Josh, 38, in early retirement because of mental health problems, lives with his partner. Has a child, placed in foster-care.
> Erving, 46, in early retirement because of multiple sclerosis, used to work in the Netherlands as a salesman.

The first heterotopia consists of media-users with chronic diseases. Age and gender seem less important than their early retirement, which gives some of them more time at home. The six individuals in this heterotopia are heavy media-users. Their media-use is partly defined by their health situation, but differs from person to person. Niels's case, which opened this chapter, is extreme in how his chronic brain damage has determined much of his media-use. Niels did not complete his university studies, but works as a part-time research assistant on a political magazine. He is the father of nine-year-old twin sons, and is a football fan. He is a blogger and content-producer on social-network media such as Facebook, blogging on politics and producing media content on various platforms (TV, internet, print media) on issues to do with chronic brain disease. In brief, Niels is an extremely active media-user and media-producer, both at work and at home. Niels uses media as his memory device. Digital media allow him to structure and archive other media. Thus, although he cannot find his cookbooks in the kitchen, when he googles, he can find many recipe websites on the internet and many websites on brain food.

Niels's case in some ways parallels that of Birger (58), an engineer with a spinal necrosis. Birger works from home as a self-employed engineer, specialising in developing materials for spinal surgery. He has constructed his iGoogle home-page around health science news, extracted from academic journals in English and German. He uploads political comments on handicap issues. Birger has cooked for his family every day for many years, but never uses recipes or online newsletters and never googles for food: he has a collection of cookbooks, magazines and handwritten recipes that he uses for his traditional Danish cooking (potatoes, gravy, meatballs), cooking that he learned at home as a boy. Although he has given his children cookbooks, for him, as for Niels, cooking and eating are something you do to survive. He has strongly negative opinions both on organic food production and on food additives, which he checks out on the internet. He always reads the nutritional information declaration and labelling on the packaging of the food he buys, and prefers Danish-origin food. Birger and Niels confirm research showing that people with chronic disease are more active in the blogosphere than others (Miller et al., 2011, p. 726, p. 732). Birger and Niels differ, however, in their food-related media-use: Birger uses print media only (cookbooks, newspapers, magazines and a selection of printed or written recipes), while Niels uses digital media only.

The third case in this heterotopia is Gitte, 63. While Niels considers himself a political consumer and his preferred websites are leftist, and while Birger considers himself a scientifically well-informed consumer and citizen but uses traditional media in relation to food and prefers Danish products and recipes, Gitte, by contrast, seems not to have any normativity in relation to media-use. She is, however, normative in relation to food.

Gitte is in early retirement. She used to work part-time in a bank. She is married, and lives in the same block as her three grown-up children. She collects print media, such as the big supermarket chains' advertising weeklies. In her kitchen, where we interviewed her, she collects these in folders:

> On a typical Monday morning – Mondays are special, because we have a cottage, and we return Sunday evening. Then I sort out the advertisement flyers that have arrived. . . . And then I collect and sort out for the woman on the third floor, she is 104 years old. . . . I look after her. . . . And then I cut the pages or simply tear them out. And then if I find some you can take part in, quizzes, competitions, I cut them out, and I cut them out if I find good offers, and then I put them here in a stack, and then I place it in this folder for good offers. The other folder is for quizzes and that one if I find something I will use later.
>
> (GK: 32–40)

Gitte is a quizzomaniac: 'This thing about winning, that's my job, you see' (GK: 490). Over the years this trait has made her an expert on consumer legislation. She knows where to find laws on the internet, and she often complains on her own and on other people's behalf. She has never lost a case in the courts system. Gitte is a political consumer, in the sense that she is well informed about consumer legislation and uses media to support her consumer politics.

Gitte uses recipes from her many print magazine weeklies, but she also uses the weeklies' websites to find recipes and food advice. Like Niels, she uses digital media as an archive for print media. She has won several cookbooks in quizzes, and she uses them. During the interview, several piles of cookbooks lay on her kitchen counter, and there were pots simmering and boiling. Gitte likes to cook and prefers good-quality food, and in her home this means avoiding convenience products.

Gitte is a media addict, but in quite another sense than Niels and Birger. She quizzes and entertains herself, and she loves to participate:

> As soon as the morning television ends, I have problems, because I have to listen to P4 radio [the most popular Danish radio channel], it is a must and I participate very much in the debate – they know me as time goes on . . . and it is a problem because I cannot do anything else. . . . It is very often I come on the air if there is something I have to comment. I do comment a lot.
>
> (GK: 42–50)

She also participates in the quizzes, but is regularly blacklisted because she phones in too often.

Like all the other respondents, Gitte likes to watch *Spise med Price* (see Leer's chapter in this volume), a popular Danish cookery show. She follows the Price brothers because they are celebrities, just as she follows other celebrities from her weeklies on TV and on the internet – food celebrities or not.

Niels, Birger and Gitte resemble the other respondents chosen for interview in that their answers to the survey showed that they were active media-users in relation to food and health. They are not far apart in age, but their media profiles differ markedly except for the common factor that they are all heavy media-users. All three are engaged and intense users of digital media, and all three confirm research findings that, despite weaker technical competencies, old and chronically ill media-users are better than younger users at selecting and using relevant digital information from reliable sources (van Deursen and van Dijk, 2011; van Deursen, 2012). The three younger men in this heterotopia are less extreme in their media-uses. While Erving is a food addict to the degree that he had stomach-reducing surgery, he is not a media addict (except for the *Spise med Price* TV show). He collects scandalous news about food fraud, unlike Josh, who knows a great deal about organic food and food regulation. Both Erving and Josh like to cook, like Torben S., who might even be placed in another category because of his fondness of cooking.

These five or six cases demonstrate how media-use is domesticated and embedded in daily practices and routines (Hollows, 2008; Hartmann, 2010; Hepp, 2013). For Niels and Birger, the point of origin of their media-use was the health issue. Gitte's, Josh's and Erving's health problems have not affected their food media-use, but have influenced the amount of time they spend using media. All six individuals belong to the taste regime of traditional Danish cuisine, with no pretension to gourmet distinctions – although Torben and Erving like to present themselves as gourmets. Although their food media-use differs, all six use media with food content to establish personal spaces of compensating heterotopias that allow them to forget their bad health. All have arranged their homes so that the PC and television are the centrepieces in the main room; similarly, for all six, their existence as media-users represents the singularity of their existence. Media-use – not least in relation to food contents – is their space of difference, of heterotopia. Torben and Erving also present cooking as a space of distinction and heterotopia, but their cooking is mainstream, and food media mentioned in the interviews (weeklies, advertisements, traditional cookbooks) are not distinctive, with one exception:

TORBEN: If we are going to have chops, I search for recipes on the internet.
INTERVIEWER: You google?
TORBEN: I go to www.opskrifter.dk (recipes.dk). Once I googled and hit the page and found it interesting. . . . I can write down ingredients in my refrigerator and search for recipes that match them.

(TS: 160)

Torben is the only one in the material who doesn't 'just google.'

To conclude: media-user patterns, which in the quantitative surveys appear similar, reveal complex individual diversity once they are investigated in the qualitative interviews. All five respondents have elaborated their own personalised media heterotopias: real and imagined extensions and networks that represent the places or spaces where they demonstrate their individuality. These are spaces for otherness, and they are often also media heterotopologies – media that reflect on the specific contents of the heterotopia. Last but not least, all five or six respondents use digital media as heterochronisms – as archives for digital and other media.

Heterotopia II: health as much more than a professional space

Rie, 36, lives with husband and two small sons in central Copenhagen, works as a nurse.

Helga, 60, single with two grown-up sons, lives in a small town in northern Jutland. Worked as a nurse for years, but after serious stress, she studied journalism and worked with hospital communication for some years. In early retirement.

Rita, 49, lives with her husband and two sons in Aarhus, works as a nurse.

Lica, 53, lives with her husband and two teenage daughters north of Copenhagen, works as a nurse.

Mette, 46, lives with her husband and three sons, doing a master's degree.

Among the respondents who accepted our invitation for interview were five nurses: four in the *Changing Boundaries* project and one in the *Multi-Trust* project. Their interest in food and health must have urged them to participate, because no other group of professionals was so eager to participate. All five nurses were female.

The individualised heterotopias are diverse in this group also, but can be described as heterotopias of health constructions. But the single spaces of heterotopias are diverse.

The youngest nurse, Rie (36), lives in Copenhagen with her two small sons and husband. Her youngest son suffers from severe allergic attacks. This dominates her media-use in relation to food. She reads about food and allergens, and buys special products for her son on German and Swedish internet sites. In fact Rie is a 'Price-runner.' Early in the interview, she states: 'I really like am so fond of advertisements and catalogues in print and online' (Rie, 54). She collects free samples and prices and buys and sells at flea and food markets.

The oldest nurse, Helga (60), does not use media heavily in relation to food. She watches TV cookery shows *(Price)* as entertainment, but seldom googles a recipe or uses food websites. What interests her is illness, cures and food production politics. She is critical of industrial pig production because of the public health issues it creates. Helga is a political consumer. Rie and Helga create heterotopias about healthier food politics in their media-uses. Rie wants to cure her son, Helga to act as an environmentalist. Both spend as little as possible on food.

The other three nurses also demonstrate diverse heterotopias in their media-uses. They have three media spaces concerning food: health and fitness (Rita), organic and mentally healthy lifestyle (Lica), and cooking and food as fun (Mette).

The three nurses thus construct three different heterotopias, but they all have in common the utopia of a healthier lifestyle.

Rita, 49, from Aarhus, Jutland, lives with her husband and two teenage sons and works with food and lifestyle in relation to her patients and clients. She is familiar with all the official Danish websites giving food and health advice – and she uses them in her own family. She was an early bird using the smartphone, and is dependent on her easy access to the internet and her daily postings on Facebook on the theme of food and fitness. She takes turns cooking with other family members, but she decides most of the recipes. She rarely googles recipes, but subscribes to them from specific healthy-eating websites run by official health authorities. Apps on her iPhone guide her to healthy, fat-free food in the supermarket, but she buys vitamins, proteins and other dietary supplements that are prohibited in Denmark on foreign websites. She subscribes to print fitness magazines to get access to recipes and advice on their websites, and she is addicted to quizzes and games on the internet on healthy food and fitness.

Rita has begun to blog about fitness and healthy, low-fat food. She explains that there can be conflict in the family because her husband collects print cookbooks, buys gourmet magazines and cooks gourmet food at the weekends. She disapproves of this interest in taste and quality and of the unhealthy food he cooks.

Nurse Lica (53), from the outskirts of Copenhagen, with two teenage daughters and a husband, is also interested in health and fitness. Like Rita, she uses some of the official health websites. She is, however, an organic food consumer, and buys food either from *www.aarstiderne.com,* which delivers a food box once a week, or from organic farm shops on the web. Lica does not watch television. She exercises in the evenings. Nevertheless she knows TV programmes such as *Bonderøven* (a Danish lifestyle series about retro living in the countryside) and *River Cottage,* and Swedish formats on organic cooking and gardening. Rita and Lica have different media-user patterns and construct very different media heterotopias. When we compare their answers to the surveys, Lica downplays her media-use in relation to food, while Rita emphasises her media activities. Their moral economy and their normativity towards media-use differ (Silverstone et al., 1992), with the result that their media heterotopias seem to differ more than they probably do.

The third nurse, Mette (46), with three sons and a husband, lives in the countryside on the North Sea coast near a harbour where fish is landed and auctioned. She commutes for two hours by car every day and is doing a master's degree in health management. She teaches fitness in her village and in addition is a hobby gourmet. She receives email newsletters from several food sites, and advertising weeklies from a discount supermarket that stocks many organic food products. She often buys cookbooks, but when it comes to recipes, she googles: 'I do not subscribe to food magazines. I can search on the internet if I need a recipe that I do not have, but I do buy many cookbooks' (Mette, 178).

Mette knows many recipe websites by name and web address. She compares them, compares ratings and even compares recipes in cookbooks and then cooks her own version. She enjoys most food formats on TV. On her holidays, she travels to Italy to taste and buy food. Mette buys meat and fish in her village, but buys most other food products on the internet and gets it delivered. She buys wine, oil and other things from Italian websites, and has strong opinions about 'food of poor quality.' Of all the interviewees, she is the one who buys most on the internet, and who uses most leisure time on food-related activities and on media food. When asked what her favourite media is, she answers: 'Oh it must definitely be Google, it contains almost everything' (342). For Mette, cooking and media-use in relation to food constitute her personal oasis, where she can forget her stress and the long hours in the car. Mette uses media and food as a heterotopia in a very literary sense of the word: as another space, or a utopian extension of everyday life. For Mette, heterotopia and utopia seem to converge.

All the nurses in this study were interested in health and food, but their versions of health differed, as did their media-uses in relation to health. While Mette's heterotopia of media and food is also her vision of a food utopia, Rita, on the other hand, is not far from regarding media as a kind of utopia in her life, because media technologies help her in her search for the perfectly healthy life. Paradoxically, this makes her a mainstream media-user, using official sites that are affirmative of existing society. Heterotopia and utopia may partly converge here, but the term heterotopia allows us to see media food as an extension of and a space for diversity in media and food practices, as well as sites for homogeneity in the imaginations of media food.

Mette regards cooking and food-related media-use as the oasis in her busy life. For her, food as media content is both utopia and heterotopia, as well as a practical device for ordering and becoming informed about available products – just as it is for Johnston and Baumann's foodies (2010).

Heterotopia III: foodies

Mette, 46, see above, Heterotopia II.
Torben S., 53, see above, Heterotopia I.
Ditte, 29, university degree in music, single, works with music pedagogics.
Jens K., 63, married, grown-up children, cycles hours each day.
René, 30, university degree in sociology, lives with his partner in central Copenhagen.
Villads, 35, university degree, works in the army administration and as a fitness instructor, married with a small son.
Jørgen, 42, self-employed, divorced, two children, Copenhagen area.

Mette does not regard herself as a foodie. She likes to cook, and enjoys food as a way of releasing stress rather than an area in which she can refine her taste distinctions (Johnston and Baumann, 2010).

The same is true for Ditte (29), a single musician and educationalist. She likes to cook and eat, but for her, food in media such as TV is pure entertainment. In print media and on the internet, she spends little time looking for food content. When she cooks, she uses cookbooks and her deceased mother's recipes, and cooking this organic food is a nostalgic, even therapeutic practice for her.

Two of the respondents, Torben and Jørgen, describe themselves as foodies and present themselves as gourmets, but they are very different in their food and media preferences. Torben (53) is university-educated and lives with his partner in central Copenhagen. His media-use both at home (he runs a specialised travel agency) and at work is extensive. He googles recipes, or looks for them in his numerous traditional Danish cookbooks or in the food magazine to which he subscribes, but he often makes his own version of the recipes. He enjoys cooking, and also enjoys TV cooking shows in the early morning and in the evening. He follows most of these, but has never tried any of the recipes available online. He prefers his traditional Danish cookbooks, and does not even know if the food magazine he subscribes to has a website with recipes. Torben is a traditionalist in regard to food as well as food-related media-use. He enjoys cooking more than he does eating the results. He seldom eats out, but reads food and restaurant reviews on *www.politiken.dk* (the newspaper website, with extensive culture, food and lifestyle content). He also likes to watch the TV chefs to relax and for entertainment. Torben's media-food uses are mainstream, but heavy. He does not appear to have constructed either a heterotopia or a utopia.

Jørgen differs from Torben in that he is divorced and lives with his part-time children in the suburbs. He has a professional training, and owns his own small business. He asks to be interviewed in a health food café in a mall, and several times stresses health issues. But really he is an all-year grill enthusiast. He does not watch TV cooking shows in the evening, but sometimes he gets tips from morning TV with easy-cook recipes. He has numerous grill and traditional Danish cookbooks, but mostly, 'I just google.' When he invites guests, however, 'it is better with a cookbook.' Torben and Jørgen often use the vocabulary of foodies and gastronomes when they talk about quality and taste:

> I cook every day, some things are more demanding than others, but I do not buy a pizza on my way home. . . . I really am interested in food . . . I really enjoy cooking . . . we do a lot of grill-food.
>
> (Jørgen, 98–102)

> It is very relaxing to cook when I come home after work. If we get visitors, I like to look through my cookbooks, you can find inspiration on the internet but books are better.
>
> (Torben, 112)

Their mainstream media-food uses make it impossible to argue that they produce heterotopic spaces in media-use or in imagined food preferences. Their cultural capital (Naccarato and LeBesco, 2012) is average, so they are 'normal' in relation to the field of taste distinctions. They are examples of how food and food-related media-use have become mainstream in the middle class.

Among these interviewees, only two males, namely René (30) and Villads (35), qualify as foodies, taking into consideration that Mette (46) is also partly a foodie – and Torben is a foodie wannabe, someone who defines him- or herself as a gourmet in media as well as in cooking practices, constructing media practices and tasting practices as authentic and exceptional:

> Quality, rarity, locality, organic, hand-made, creativity, and simplicity all work to signify specific foods as a source of distinction for those with cultural and economic capital.
>
> (Johnston and Baumann, 2010, p. 3)

As with a fashion cycle (Simmel, 1957), elite food professionals and food enthusiasts push the boundaries of what is considered daring, bold and exotic. Some (but not all) of these trends slowly filter down to mainstream eaters and are then reclassified as bland or passé by the food avant-garde (Johnston and Baumann, 2010, p. 25).

The two unambiguous foodies in this project are both male, have university degrees and are intense, wide-ranging media-food users. Both make explicit distinctions of taste in relation to media-use as well as food and cooking practices.

René (30) lives with his partner. He has a university degree in sociology and works round the clock in a market-analysis bureau. He is a heavy media-user: 'Then it is like a small ritual when I arrive at work in the morning, where work and leisure melt together' (R: 36). The media ritual consists of browsing on sports websites, elite-level running websites, private emails, food websites, digital newspapers, professional websites and digital food magazines and blogs. René cooks every day for himself and his partner. He 'likes to cook and eat, so it is something I feel great joy in. . . . I like to spend time on cooking something good and delicious' (58–60). He gets his inspiration from the internet, consumer advice and tests, newspaper reviews *(Politiken),* tests and *YouTube* videos. He googles recipes, but distinguishes between hits, weighting ratings and credibility as important factors.

René likes to stream TV cooking shows, and he likes to pretend that he is the cook. He often cooks TV chef recipes and has some of their cookbooks, but also finds recipes on the internet in the digital newspapers he trusts *(politiken.dk).* As a former elite-level runner, he is not interested in sports TV or amateur websites, which he finds 'below my ambition.' The same is true for food and recipes. Organic commodities are more interesting to him from a gastronomic point of view:

> I buy a lot organic food and I am willing to pay extra. . . . I find the products more exciting . . . and healthier, of course . . . good products that give me a nice experience cooking and eating them.
>
> (R: 238–240)

He does not read labels, and is not concerned about artificial additives: what he cares about is taste and distinctions from the average culinary capitals. His heterotopia of food media-use is one of personal combinations, which establish his unique user profile and thus comprise distinctions that are unique. According to

Johnston and Baumann (2010), such heterotopias of personal distinctions are not unusual; in fact they are typical of foodies.

Villads (35) is older than René and has a baby son. He too has a university degree, an army job and is also a former elite-level runner. Six months ago his first son was born, and that changed his life. For many years he had considered himself a foodie, alternating between high cuisine and junk food. Now he is also into organics, wanting clean food for his son. Villads and his family have just bought a new house; they bought the house to replace their kitchen, which they renovated but which is still too small for Villads's cookbook collection. Villads trusts chefs and styles (Wassim, the Fatduck) and often cooks for friends all weekend, always moving from the internet to the cookbook rather than the other way around. He does not trust the internet or Google, although he often googles an ingredient because he is highly conscious of the relation between price and quality. When he finds a good bargain, he googles on his smartphone and then orients himself in his cookbooks at home. These are Danish, English, American and Australian, and they are meat-based. Villads himself likes a boeuf, but for his son he goes for organic and vegetarian.

Villads is a cookbook fanatic and a film and magazine fanatic. He subscribes to numerous food and film magazines. For him, trust and credibility are important, and he has more trust in print media than in the internet. His worst fantasy is that his son – or he himself – might become overweight. They like to lie on the couch in the evenings and eat sweets together while watching TV. And when it comes to it:

> I almost prefer junk food, pizza and burgers, not every day of course, but that is what I like best. Luckily we have the best organic pizzeria in Copenhagen just around the corner. Now we cook five out of seven days a week.
>
> (V: 286)

In every respect, Villads is a foodie as defined by Johnston and Baumann (2010). He is also a media-user with a wide range and variety of media-use, from cookbooks to *Endomondo,* from the app *Karolines Køkken* (traditional Danish dairy kitchen) to Hester Blumenthal's cooking shows on TV.

Villads is a foodie combining diverse coexistent heterotopias in his media and food practices. He is also an omnivore (Peterson and Kern, 1996) in respect to media and food. He appreciates both high cuisine and junk food, and uses international gourmet cookbooks and *www.pricerunner.com* to inform himself and keep up to date. Villads's heterotopia is thus unique, yet at the same time typical in its uniqueness. His heterotopia is under constant construction in regard to creating new distinctions in media-uses as well as food preferences. He is on the move towards new horizons.

Heterotopia: coexistent, combined and connected

The interviews show that in the search for new taste experiences and distinctions, the interviewees often refer to the media even when they have not been asked about media-use. They mention reading books, blogs, reviews, newspapers, recipes, watching films and television, and searching on the internet – the more, the merrier. A broad use of food media often signals a broad competence across different taste regimes. The food omnivore is usually also a media omnivore. Food omnivores were a minority, however, in both the representative surveys and the qualitative interviews.

Nevertheless, the majority is no homogeneous mass – which is why it has been difficult to single out three spaces of heterotopias. Most interviewees use media food to create diverse heterotopias, to dig into heterotopologies and, not least, to use the media as heterochronisms. Many use the internet as an archive for food content. Everyone but one googles, but nobody 'just googles.' All perform personal and complex media routines regarding food. Media-user patterns thus draw pictures of different and personalised taste regimes in relation to food. The chapter has demonstrated, however, that a mainstream media-food user does exist, and that the fluffy term 'heterotopia' makes possible an understanding of diversity in the details, as well as similarities on a structural level in media-user patterns and food preferences.

Already in 2006, Jacquie Reilly stated in 'The Impact of the Media on Food Choice' that it is not possible to trace a direct effect from media content in, for instance, popular TV cooking shows or big official campaigns. Media-users construct their own media palette; they perceive chefs as trustworthier than official advice, and cookbooks as more credible than Google. In the period 2010–14, her statement seems even truer. Media-use has become so diverse that the pattern now seems to be one of difference rather than sameness. Hegemonic sameness still exists in food preferences as well as in media-use. But when the two fields are combined, only the few price conscious consumers follow mainstream food and media consumption. Even education or profession do not suggest the same hegemonic field. Taste is performed as a distinction, both in relation to food and recipes and in relation to media. This pattern of difference makes Foucault's conception of heterotopia even more relevant today than in the 1960s. In a world of individualisation and fragmentation, media choices and food choices seem to duplicate the heterotopias possible in ways that do not replicate traditional segmentations.

Notes

1 Interview citations throughout this chapter include interviewee's initials and the number of the transcription line.
2 See also Halkier in this volume.
3 *Changing Borders,* funded by the independent Danish Research Council (FKK) 2008–11.
4 *Multi-Trust: Trust in Organics,* funded by the Danish Ministry of Food GDP 2011–14.

References

Ang, I., 1996. *Living Room Wars: Rethinking Media Audiences for a Postmodern World.* London: Routledge.

Appadurai, A., 1988. How to Make a National Cuisine. *Comparative Studies in Society and History*, 30(1), pp. 3–24.

Ashley, B., J. Hollows, S. Jones and B. Taylor, 2004. *Food and Cultural Studies.* London: Routledge.

Beardsworth, A. and T. Keil, 1997. *Sociology on the Menu.* London: Routledge.

Bell, D. and J. Hollows, 2005. *Ordinary Lifestyles.* Nottingham Trent: Open University Press.

Bell, D. and J. Hollows, 2011. From *River Cottage* to *Chicken Run:* Hugh Fearnley-Whittingstall and the Class Politics of Ethical Consumption. *Celebrity Studies,* 2(2), pp. 178–191.

Bird, E.E., 2003. *The Audience in Everyday Life.* London: Routledge.

Bonner, F., 2005. Whose lifestyle is it anyway? In: D. Bell and J. Hollows, eds. *Ordinary Lifestyles.* Maidenhead: Open University Press, pp. 35–46.

Bonner, F., 2009. Early Multi-Platforming. *Media History,* 15(3), pp. 345–358.

Bourdieu, P., 1984 (1979). *Distinction.* London: Routledge.

Bräuchler, B. and J. Postill, eds., 2010. Introduction. *Theorising Media as Practice.* Oxford: Berghahn Books, pp. 1–34.

Butler, J., 1990. *Gender Trouble.* London: Routledge.

Castells, M., 1996. *The Rise of the Network Society.* Oxford: Blackwell.

Couldry, N., 2004. Theorising Media as Practice. *Social Semiotics,* 14(2), pp. 115–132.

Couldry, N., 2008. Mediatisation or Mediation? *New Media and Society,* 10(3), pp. 373–391.

Couldry, N., 2009. Does 'The Media' Have a Future? *European Journal of Communication,* 24(4), pp. 437–449.

Couldry, N., 2010. Theorising media as practice. In. B. Bräuchler and J. Postill, eds. *Theorising Media and Practice.* Anthropology of Media, 4. Oxford: Berghahn Books, pp. 35–54.

Couldry, N., 2012. *Media, Society, World: Social Theory and Digital Media Practice.* Cambridge: Polity.

Couldry, N., S. Livingstone and T. Markham, 2007. *Media Consumption and Public Engagement.* Basingstoke: Palgrave Macmillan.

Counihan, C., 1999. *The Anthropology of Food and Body.* London: Routledge.

Counihan, C. and P.v. Esterik, 2008. *Food and Culture, 2. Ed.* London: Routledge. http://changingborders.au.dk/wp-content/uploads/2012/05/The-Media-Menus-of-Danish-Internet-Users-2009.pdf.

DeSoucey, M., 2013. Food. In: Oxfordbibliographies.com.

Douglas, M., 1972. Deciphering a Meal. *Daedalus,* 101(1), pp. 61–81.

Finnemann, N.O., 2005. *Internettet I mediehistorisk perspektiv.* København: Samfundslitteratur.

Finnemann, N.O., P. Jauert, J.L. Jensen, K.K. Povlsen and A.S. Søresensen, 2012. The Media Menus of Danish Internet Users. Internet publication http://changingborders.au.dk/wp-content/uploads/2012/05/The-Media-Menus-of-Danish-Internet-Users-2009.pdf.

Fischler, C. 1988. Food, Self and Identity. *Social Science Information,* 27(2), pp. 275–92.

Flyvbjerg, B., 2006. Five Misunderstandings about Case-Study Research. *Qualitative Inquiry,* 12(2), pp. 219–245.

Foucault, M., 1978. *The History of Sexuality I.* Harmondsworth: Penguin.

Foucault, M., 1997 (1967). Of other spaces: Utopias and heterotopias. In: Neil Leach, ed. *Rethinking Architecture.* London: Routledge, pp. 330–336.

Goffmann, E., 1957. *The Representation of Self.* Harmondsworth: Penguin.

Goody, J., 1977. *The Domestication of the Savage Mind.* Cambridge: Cambridge University Press.

Halkier, B., 2010. *Consumption Challenged: Food in Medialised Every Day Life.* Burlington: Ashgate.
Hartmann, M., 2006. The triple articulation of ICTs: Media as technological objects, symbolic environments and individual texts. In: T. Berker, M. Hartmann, Y. Punie and K. Ward, eds. *Domestication of Media and Technology.* Maidenhead: Open University Press, pp. 80–102.
Hartmann, M., 2010. Mediatisierung als mediation. In: M. Hartmann and A. Hepp, eds. *Die Mediatisierung der Alltagswelt.* Wiesbaden: Verlag für Sozialwissenschaften, pp. 35–48.
Hartmann, M., 2011. *Moving Around, Moving Us.* Unpublished paper delivered at the conference: Mediated Social Connections and Cultural Engagements, SDU, 7–8 February 2011.
Hepp, A., 2013. *Cultures of Mediatisation.* London: Polity.
Hepp, A., S. Hjarvard and K. Lundby, 2015. Mediatization: Theorizing the Interplay between Media, Culture and Society. *Media, Culture & Society*, 37(2), pp. 1–11.
Hjarvard, S., 2009. Samfundets Medialisering. *Nordicom-Information*, 31(1–2), pp. 5–35.
Hollows, J., 2008. *Domestic Cultures.* Maidenhead: Open University Press.
Hollows, J. and S. Jones, 2010a. 'Please Don't Try This at Home': Heston Blumenthal, Cookery TV and the Culinary Field. *Food, Culture and Society*, 13(4), pp. 521–537.
Hollows, J. and S. Jones, 2010b. 'At Least He's Doing Something': Moral Entrepreneurship and Individual Responsibility in *Jamie's Ministry of Food. European Journal of Cultural Studies*, 13(3), pp. 307–322.
Johnson, P., 2013. The Geographies of Heterotopia. *Geography Compass*, 7(11), pp. 790–803.
Johnston, J. and S. Baumann, 2010. *Foodies Democracy and Distinction in the Gourmet Landscape.* Routledge: London.
Katz, E. and T. Liebes, 1984. Once Upon a Time in Dallas. *Intermedia*, 12(3), pp. 28–32.
Krotz, F., 2007. The Meta-Process of 'Mediatization' as a Conceptual Frame. *Global Media and Communication*, 3, pp. 256–260.
Leer, J., 2014. Ma(d)skulinitet. PhD thesis, University of Copenhagen.
Levi-Strauss, C., 1966. The Culinary Trainge. *Partisan Review*, 33(4), pp. 586–95.
Livingstone, S., 2009. On the Mediation of Everything. *Journal of Communication*, 59(1), pp. 1–18.
Livingstone, S. and P. Lunt, 2014. Mediatization: An emerging paradigm for media and communication research. In: K. Lundby, ed. *Mediatization of Communication.* Handbooks of Communication Research, 21. Berlin: De Gruyter Mouton, pp. 703–723.
Lupton, D., 1996. *Food, the Body and the Self.* London: Sage.
Mennell, S., 1985. *All Manners of Food.* Urbana: University of Illinois Press.
Miller, E.A., A. Pole and C. Bateman, 2011. Variation in Health Blog Features and Elements by Gender, Occupation and Perspective. *Journal of Health Communication*, 16(7), pp. 726–749.
Miller, T., 2007. *Cultural Citizenship.* Philadelphia: Temple University Press.
Murcott, A., ed. 1983. *The Sociology of Food and Eating.* Aldershot UK: Gower.
Naccarato, P. and K. LeBesco, 2012. *Culinary Capital.* Oxford: Berg.
Neergaard, H., 2007. *Udvælgelse af Cases.* Samfundslitteratur: Roskilde.
Peterson, R.A. and R.M. Kern, 1996. Changing Highbrow Taste: From Snob to Omnivore. *American Sociological Review*, 61(5), pp. 900–907.
Povlsen, K.K., 1986. *Blikfang* (Eye Catcher). Aalborg: Aalborg Universitets Forlag.
Povlsen, K.K., 1999. *Beverly Hills 90210. Ironi, soaps og unge* (BH90210). Aarhus: Klim.

Reckwitz, A., 2002. Toward a Theory of Social Practices. *European Journal of Social Theory*, 5(2), pp. 243–63.

Reilly, J., 2006. The impact of the media on food choice. In: R. Shepherd and M. Raats, eds. *Psychology of Food Choice*. Wallingford: Cabi Publishing, pp. 201–225.

Rittenhofer, I. and K.K. Povlsen, 2015. Trust and Credibility. *Ecology and Society*, 20(1), p. 6. http://www.ecologyandsociety.org/vol20/iss1/art6/

Schroeder, K., K. Drotner, S. Kline and C. Murray, 2003. *Researching Audiences*. London: Arnold.

Schulz, W., 2004. Reconstructing Mediatization as an Analytical Concept. *European Journal of Communication*, 19(87), pp. 87–101.

Silverstone, R. and E. Hirsch, eds., 1992. *Consuming Technologies: Media and Information in Domestic Spaces*. London and New York: Routledge.

Simmel, G., 1957. Fashion. *American Journal of Sociology*, 62(6), pp. 541–558.

Thompson, J., 2013. *The Media and Modernity*. London: Polity.

Thorsø, M., K.K. Povlsen, T. Christensen, forthcoming 2016. Trust in Organics. *Food, Culture and Society*, 17(4).

van Deursen, A.J.A.M., 2012. Age and internet skills: Rethinking the obvious. In: E. Loos, L. Haddon and E. Mante-Meijer, eds. *Generational Use of New Media*. Surrey: Ashgate, pp. 171–184.

van Deursen, A.J.A.M and J.A.G.M van Dijk, 2011. Internet Skills and the Digital Divide. *New Media and Society*, 13(6), pp. 893–911.Warde, A., 1997. *Consumption, Food and Taste*. London: Sage.

Warde, A., 2005. Consumption and Theories of Practice. *Journal of Consumer Culture*, 5, pp. 131–153.

Yin, R.K., 2009. *Case Study Research*. London: Sage.

9 Everyday mothering and the media food 'soup'

Comparing contested food and mothering across genres in two different social contexts

Bente Halkier

'Mum' is still taking the lion's share of food-related division of labour in families, also in the context of the Scandinavian welfare state, although some changes can be registered (Holm, 2013). Thus, the family activities of food provisioning, cooking and eating, which span the material and corporal taken-for-granted procedures and the symbolic importance of social and cultural identification, are closely associated with motherhood.

At the same time, media discourses on food in Denmark have led to an increasing questioning of everyday food routines in society. Societal consequences of food habits in relation to e.g. climate, health, risk and quality are being framed as problematic and the individual or family framed as responsible for contributing to solutions to these collective problems by way of changing their own routines. These kinds of normative discourses appear across different types of media food genres, from recipe resources and marketing framing, over television and lifestyle magazine entertainment, to social media posting and interaction – and not just in the genre of public communication campaigns. Furthermore, due to the media development (Bennett and Iyengar, 2008) and the mediatisation of culture (Hepp, 2012), everyday life as a context is ever more media-saturated (Couldry, 2004), meaning that it is difficult to avoid media discourses, hence also the cross-genre discourses (Bjur et al., 2013) of the 'soup' of media food. This way, food activities in everyday life become culturally and normatively contested (Halkier, 2010). Unnoticed food routines potentially become noticed and already noticed food signals potentially become re-framed. Hence, acting as mother through food activities also become potentially questioned, negotiable and re-framed. The question is how do particular kinds of representation and framing of food in media discourses interplay with everyday activities among particular mothers?

This chapter compares 'mothering' through food practicing among two different groups of Danish women where their food practices are related to two different media food genres. The one study concerns Danish Pakistani women who as minority Danes have been particularly targeted by public healthier diet communication and campaigns. The other study concerns majority Danish female readers of a particular lifestyle magazine which, among other things, covers gardening and cooking. The patterns of 'mothering' come from these two qualitative empirical studies.

The chapter falls in four parts. First, the theoretical perspective on food and media food is presented, being a practice theoretical approach, where media discourses are seen as resources that are being domesticated. Second, the methodological design of the two empirical studies of food practicing in mediatised everyday contexts is shortly presented and an argumentation for the validity of comparing the empirical patterns across the two different social contexts as well as media food genres is given. Third, a number of examples on empirical patterns in using media food framings for mothering through food are presented. The focus is on differences in cultural hybridisation in food practicing related to the media food genre, and on similarities in mothering through food across connections with different media food genres. Fourth, the conclusion ties back to the current media development.

A practice theoretical approach to food practicing, mothering and media food

In my understanding, a practice theoretical approach is a particular reading of an assembly of theoretical elements from, among others, the early Pierre Bourdieu's (1990) concepts of habitus and field, Judith Butler's (1990) understanding of performance, the early Anthony Giddens' (1984) structuration theory, and the late Michel Foucault's (1978) thinking about social regulation of bodies through discourses. The shared assumptions among these theoreticians about how social action is carried out and carried through are central in practice theories. Thus, a practice theoretical approach primarily conceptualises the organisation and accomplishment of mundane performativity in micro-contexts.

A practice theoretical approach is neither based on epistemological individualism or structuralism (Reckwitz, 2002). The unit of analysis is practices and performances (ways of practicing), and neither the individual mother or cook, nor the cultural scripts of mothers and cooks. The concept of social practices is used to cover any kind of ordinary activity, such as eating practices, parenting practices and citizen practices. Every practice is organised as a nexus (Schatzki, 2002) by a number of equally important elements, such as meanings, materials and competences (Shove et al., 2012), or understandings, procedures and engagements (Warde, 2005). I am drawing upon the latter version in my empirical work. Hence, the concrete activities in practicing something – eating or mothering – always cover both doings and sayings (Schatzki, 2002). All micro-contexts are understood as intersected by a plurality of such different practices. People are thus more often than not doing several overlapping activities, which therefore contain overlapping and ambiguous social expectations for how to perform as e.g. good cook, responsible mother or engaged citizen.

Practicing motherhood is thus one of these overlapping activities, and I refer to it as 'mothering'. By using the term 'mothering', I place my argumentation in the part of the literature on food, family and gender that underlines the importance of the performative character of 'doing' motherhood (e.g. Clarke, 2004; Mitchell and Green, 2002; Molander, 2011; Moloney and Fenstermaker, 2002; Skeggs, 1997).

This literature dovetails with a practice theoretical approach through the shared performativity perspective. In this manner, doing mothering is not a social category with fixed content, pre-formed by structures or values. Rather, mothering is continuously constructed, negotiated and accomplished as social category. Mothering is done, re-done and slightly differently done. Mothering is adapted and experimented with among mothering practitioners, and negotiated in relation to what constitutes expectable and acceptable mothering conduct. Meanings and interpretations in these negotiations are potentially connected with media food representations of mothering (Marshall et al., 2013).

A long line of empirical research has shown that food practices constitute an important part of mothering processes (e.g. Anving and Thorsted, 2010; Blake et al., 2009; Bugge and Almås, 2006; DeVault, 1994; Jabs and Devine, 2006; James and Curtis, 2010; Moisio et al., 2004; Molander, 2011; Murcott, 1983; Ristovski-Slijepcevic et al., 2010). So, when food practices become potentially contested via media food discourses, mothering through food becomes a normative category (Coveney, 2000; Johansson and Ossiansson, 2012; Thomson et al., 2012), and 'appropriate' mothering becomes part of an enlarged pool of overlapping activities and normative expectations. Overlapping normative expectations in everyday life can be seen as somewhat parallel to the Foucauldian notion of heterotopias, places of 'otherness' where 'real' places are both represented, contested and inverted (Foucault, 1984). Later, this concept has been extended to cover not just places, but also e.g. cultural practices, relationships and discourses. Heterotopia can thus be seen as e.g. practices or discourses which blend normality and othering (Meininger, 2013, pp. 31–32).

The media food discourses about different societal issues such as quality, risk, health and environment form part of questioning and potentially contesting mothering through food in everyday life. From a practice theoretical perspective, media discourses are primarily linked to the representational dimensions of practices and performances, such as understandings of food and mothering, engagements in food and mothering, and normative categories for negotiations about food and mothering. Another point in a practice theoretical approach to media use is that the present media saturation in everyday life makes it difficult for researchers to see and analyse media use as a discrete activity or separate practice in itself (Couldry, 2004, p. 125); rather, it overlaps and is embedded with all sorts of other practices. This way, practice theoretical approaches to media use share conceptual assumptions with 'non-media-centric' approaches to media use (Pink and Mackley, 2013, p. 681). Therefore, I understand media food discourses as symbolic resources for the food practicing and the mothering through food performing, along the lines of seeing resources and rules helping produce and condition agency in Giddens' (1984, pp. 16–28) structuration theory (see also Keller and Halkier, 2014). Thus, media food discourses are not necessarily 'privileged representations of the world' (Bird, 2010, p. 87), but rather pull and push resources to be appropriated, negotiated and re-framed by mothers as food practitioners in their everyday lives.

This is an understanding of media in everyday life that comes close to the understanding embedded in the concept of domestication of media (Hartmann, 2013, pp. 43–46; Silverstone, 1994, pp. 122–131). Domestication refers to the processes by which everyday actors integrate media into their lives and make them 'their own'. These domesticating processes cover the material elements as well as the representations of media, the so-called 'double articulation' (Silverstone, 1994, p. 122). Integrating the representational and the material resonates with practice theoretical reasoning (Warde, 2005, p. 133), and current media research underlines exactly the combination of the symbolic and the material in arguments on domestication (Hartmann, 2013, p. 43). In her use of the concept of domestication, Hartmann offers an understanding of media use analysis like Couldry (2004): Media uses cannot be separated analytically from the field of practices into which they are embedded (Hartmann, 2009, p. 422). The current media development underlines the potential contingency in the domestication processes of media food, because of the multiplicity of different media discourses in and across different genres through different platforms (Bennet and Iyengar, 2008). How different types of media food become domesticated thus depends upon the constellations of practices and performances in everyday life mothering where media food forms part of the resources. Furthermore, it can be difficult to separate out the domestication of meanings represented in one type of media food from the total mesh of media food representations, due to the tendency of media use taking place across genres in media-saturated everyday lives (Bjur et al., 2013).

Mothering through food and two media food genres

The qualitative empirical studies of mothering through food practices upon which this chapter is based were situated in relation to two different kinds of media food genres. Media food can take the genre of a practical resource from which food consumers can draw out what they need in terms of e.g. information, opinion, or organising device. This has been one of the characteristics of traditional cookbooks and women's magazines for many years (Floyd and Forster, 2003; Hermes, 1995; Martens and Scott, 2005; Warde, 1997, pp. 59–67). Media food can also take the genre of discourses and aesthetics for social identification processes, which is particularly distinctive in relation to lifestyle magazines (Hermes, 1995; Povlsen, 2007) but also the case for particular coffee-table cookbooks (Hollows, 2003), television shows (Ashley et al., 2004, pp. 171–185), websites, social media etc. Media food can take the genre of entertainment in many different media forms such as e.g. magazines (Hermes, 1995) and television programmes (O'Sullivan, 2005). Media food has for a long time taken the genre of marketing via commercials (Peattie and Peattie, 2003), and this genre of media food now tends to diffuse into social media forms such as apps for mobiles and Facebook (Montgomery and Chester, 2009). Finally, media food takes the genre of public communication campaigns (Windahl and Signitzer, 2009), where the media discourses on food are framed and targeted to persuade and involve groups of people to relate actively to particular messages. Again, this media food genre can take different mediated forms from the traditional folder, poster and national advertisements to documentary and entertaining television programmes and new social media. Furthermore, as a consequence of the media saturation of everyday life (Couldry, 2004) and the tendency to cross-genre media use (Bjur et al., 2013), degrees of overlap, blending and blurring of the ways in which different media food genres get drawn upon and domesticated is expected.

The first study was called 'Network communication and changes in food practices',[1] and it was a qualitative study of food habits and the handling of official communication on healthier food among Pakistani Danes. The general tendency of the discourses and visual representations of the different health campaigns is to frame food as a tool for change, to frame the citizens as only partly knowledgeable and competent (Eden, 2009) but individually responsible or responsible in the family. The 19 Pakistani Danes in the sample varied according to gender, age (15–65), and education (with and without high school degree); whether participants were born in Denmark or Pakistan; whether a person in the family had been diagnosed with type 2 diabetes; and whether participants worked in the health sector. The qualitative data material was produced by a combination of individual interviewing (Holstein and Gubrium, 2003; Spradley, 1979) with the main cook in the family, family interviews and group interviews (Frey and Fontana, 1993), participant observation at visits and parties (Hammersley and Atkinson, 1995), and auto-photography on everyday meals and weekend meals (Heisley and Levy, 1991).

The second study was called 'Cooking in Medialised society',[2] and it was a qualitative study of cooking practices, uses of the magazine *Isabellas* and constructions of cooking from scratch among female readers and their social network. The general tendency of the discourses and visual representations of the magazine on food is a celebration of ordinary everyday life activities traditionally associated with domestic femininity (Hollows, 2007), such as cooking homemade meals from scratch, baking and preserving. The 17 women in the sample varied in terms of age (20s, 30s and 40s), education level (with and without high school exams), family status (with and without children), and place of living (city, suburb and village). The qualitative data material was produced by three methods: repeated individual interviewing with the female subscriber (Atkinson, 1998; Holstein and Gubrium, 2003; Spradley, 1979), auto-photography on an everyday life meal (Heisley and Levy, 1991; Hurdley, 2007), and focus groups with the female subscriber and members of her social network (Barbour, 2007; Puchta and Potter, 2005).[3]

It is analytically valid to compare their empirical patterns on mothering through food and media food, in spite of the differences of the two studies in terms of social background of the participants and genre of media food precisely because both case-studies are examples of mothering through food, and mothering through food in the context of food contesting media discourses. Comparison is analytically sound if it takes place based on categories that are in common across different groups of participants or cases (Silverman, 2006; Thomas, 2011), which the categories of e.g. 'mothering', 'good food' and 'inappropriate food' are across the two case-studies. Second, comparison across the two different media food genres is analytically sound in spite of the specificity of each of the media food genres, because both media-genres contain 'food contesting discourses'. Although the media food in question worked as a representational sounding board in both studies, wording, phrasing and imagery in both cases reflect 'food contestation'.

Mothering in relation to media food

Performing mothering through food under the impression of different genres of media food in the big mesh of media representations of food is on the one hand full of differences and variations. On the other hand, there are some striking similarities across the two more specific genres of media food and the two different groups of media users in their ways of performing mothering through food. In this section, I will first present some empirical patterns that show some of the differences between mothering which seem to potentially be related to each specific media food genre. Second, I will present some empirical patterns that highlight some of the similarities in mothering through food in spite of the connections with as different media food genres as public health campaigns and women's lifestyle magazines.

The media contested food in relation to the public health campaigns

On top of food practices overlapping with many other practices in everyday life, food practices in themselves can also be compound. What it means is that cooking and eating practices among specific food practitioners cover activities, materials, understandings and procedures that have otherwise been seen culturally as belonging to separate repertoires or social groups. When families eat, they eat meals, food and snacks that are typically a mixture of unhealthy and healthy food. When mothers cook for their families, they can cook and/or serve a mixture of Danish, Italian and Pakistani food elements and meals. Culturally hybridised food practices are not only constellations and mixes of ethnically compound activities. There are many different social and cultural dynamics which contribute to the hybridisation of food practices. In a number of contexts, also the Danish, the American food regime has become globalised in the last 30–40 years (Ritzer, 2008), and likewise, the food tastes of consumers have become more internationalised, although in a domesticated way (Caldwell, 2004; Holm, 1996; Warde et al., 1999). The implications are e.g. a growth in the consumption of semi-produced food products, fast-food, take-away and sweet drinks, and a growing variation in culinary repertories being drawn upon in the provision of meals, such as Middle-East, Chinese, Italian etc., together with Danish.

In the study on the health campaign genre of media food, the specific hybridised ways of providing food and eating among the Pakistani Danish mothers cover three sets of cultural categories of food, namely Danish and Pakistani food, good food and bad food, and healthy and unhealthy food.[4] Pakistani food refers often to evening meals, consisting in spicy oil-roasted masala, rice and chapatti, to guest food such as fried snacks, and to sweet heavy desserts. Danish food refers to boiled food where the food elements are separated, to breakfast products such as muesli, to rye bread for lunch, and to specific meals such as meatballs and lasagne.[5] A part of what is considered Danish food is being made into 'good food' by making it more Pakistani as the following example shows:[6]

MARIA: 'Lasagne is also something that is full of flavouring, it tastes okay.'
SADA: 'But we use our own spices.'

In this manner, the quote also illustrates that Pakistani food is associated with 'good food' elements, such as richness in taste, identification with what we eat, childhood memories and homemade food. Rushy expresses it in this manner:

> My mother for example, when I think about childhood I often think Sunday, ah, I really feel like this chickpea bread she makes, and it's chickpea flour with all sorts of green and chilli and everything she puts in. And then lots of Lurpak butter on top of it and then you eat it together with a drink made of natural yogurt. It's just . . . then I go straight back into childhood. That's not something I do myself.

At the same time, parts of Pakistani food are also understood as 'bad food', associated with 'unhealthy food', which is an example of a parallel framing of the public health campaigns. The mothers refer particularly to issues of using too much oil and too much sugar. In the following, Solejma comments on the traditional way vegetables are used in Pakistani meals, by being roasted on a pan in oil to become part of the masala: 'There are a lot of healthy things, where when we are using them, we make them unhealthy.' On the other hand, Pakistani food is also associated with 'healthy food', for example by having meals without meat and by using many vegetables, beans and lentils.

The construction of Danish food is as compound as the Pakistani food. In some situations Danish food equals 'bad food', meaning with no flavours, inappropriate to eat and too much fast-food and take-away, which again is associated with 'unhealthy food'. Ayshah expresses it like this: 'Yeah, but it doesn't mean it's sick, but it just means it's Danish, it's boring, you know. It doesn't taste of anything.' The use of the connotation 'sick' in the quote relates to another category used among the mothers, which relates to the public campaign and information elements. Boiled food without spices is called 'food for the sick' and associated both with food recommended to diabetes patients and to Danish food more generally. On the other hand, parts of Danish food are also understood as 'healthy food', which again is categorised as 'good food'. Sada gives an example of how she understands Danish food as healthier food in the following:

> To me, healthy food is limited use of fat, and lots of fruit and vegetables made in the right way. Which I don't always think they are . . . yeah, it's vegetables, and if they get roasted in oil and such. And that's part of our traditional food. If only they were boiled instead. But there are many who say, no, we just can't eat that.

Thus, the food the mothers are cooking and eating with their families is culturally intersected by categories of Pakistani, Danish, good, bad, healthy and unhealthy. But unhealthy and healthy food becomes clearly enacted in these hybridised food activities of the mothers, and with clear parallels to the content of the framing of the Danish public health campaigns. Interestingly, this happened across social differences among the mothers, and it happened early on in interviews, before the issue of health was ever introduced. In the same context however, the content of the Danish public health campaigns targeted specifically towards Pakistani Danes get criticised for inappropriate normative suggestions for changing habits. See for example this exchange between an aunt and her niece:

RUSHY: '. . . and then this sort of dietician turns up and tells you about rye bread and dairy products, mayonnaise and tartare sauce and such things.'
ISHIITA: 'We can't really use that for anything.'

So, parts of the representations of food in the public health campaign genre of media food appear to have been domesticated – adapted, negotiated and made their own – among the Pakistani Danish mothers in two ways. First, the discourses of the health campaign genre are being used as a resource for appropriation in procedures for how to cook and eat in different food contexts and how to talk about this. This appropriation is, however, at the same time negotiable in relation to two other normative sets of categories, good versus bad food and Pakistani versus Danish food. Second, the discourses of the health campaign genre seem to be made a part of everyday life food practicing as a dynamic for ambivalent social disciplining. Serving healthier food for the family is constructed as a proper and normal thing to make one-self and others do, but it is at the same time constructed as food procedures which can be in conflict with other and more appropriate normative expectations of food performances as part of mothering, such as proper taste of the meal or what is usually eaten at particular meals.

The media contested food in relation to the lifestyle magazine

In the study of the lifestyle magazine genre of media food, the specific hybridised ways of providing food and eating among the magazine reading mothers also cover three sets of cultural categories of food, but slightly different ones. The distinction between good and bad food is in common with the health campaign project, but framed as 'good cooking' and 'bad cooking'. The other two pairs of distinctions are 'cooking from scratch' and 'not cooking from scratch', and 'do-able cooking' and 'unrealistic cooking'.

The category of 'good cooking' is connected with a large variety of social, cultural and normative associations. The mothers refer to knowing your craft, having a feel for the result, knowing how to handle raw materials, keeping good hygiene, taking the necessary time, using good raw materials, cooking meals from scratch, producing nutritious meals, being able to improvise, being able to 'time' the different procedures, pleasing those who are eating, producing aesthetic meals, and being able to cater for different eating-contexts. A number of these elements are defining elements in the variations of mothering through food, which are presented in the next section of the chapter, dealing with the similarities across the two studies. Furthermore, 'cooking from scratch' is one of the strong associations with 'good cooking', seen in several other studies (e.g. Bugge and Almås, 2006; Moisio et al., 2004; Short, 2006). In the following, Dorte expresses this association:

> But you know, we both like to make these homemade things, I make stewed strawberry[7] or . . . jam, we make jam once in a while, but you know, of course we also buy jam, but you know . . . I like that thing about making things from scratch, you can see what kind of things you've got your hands in, right, and proper raw materials and such things.

Pia explains how cooking from scratch is a taken-for-granted association with proper food, coming from her learning to cook in childhood from her mum: 'But you know, I've always baked, made cakes and . . . you know such things, I've always baked buns. You know, it's always been like this, if we wanted a cake, we could bake cookies or whatever it was.' This referring to 'good cooking' as 'cooking from scratch' is one of the noticeable elements in the visual aesthetics as well as the discourses in the lifestyle magazine of *Isabellas,* where baking, preserving, putting together dinner meals and learning to cook specific complicated meals take up prominent space on the pages.

However, the mothers also question the normative expectations around cooking homemade meals. 'Cooking from scratch' becomes associated with 'unrealistic cooking', which makes mothering through food less 'do-able'[8] in everyday life. Thea puts it like this: 'So a home-made soup is completely made from scratch, but preferably in an easy way, not like having to boil a chicken for 3 hours (laughs).' Across the three network focus groups with women in different ages, the mothers relate 'unrealistic cooking' directly with the representations of food in the lifestyle magazine *Isabellas*. Here is an example from the group of women in their 30s:

BRITTA: 'Hmm, it's a bit like this, you could dream yourself away, imagine if I made it, how delicious would that be . . . but there is a 1:8 chance of that happening.'

CONNIE: 'But you would make these, wouldn't you? You would make these buns where there is a little . . .'

BRITTA: '. . . ribbon round the middle with name on . . .'

ANJA: '. . . in a good year, yes!' (laughs)

Thus, 'cooking from scratch' as 'good cooking' gets negotiated with 'do-able cooking'. One of the mothers, Tilde, puts it very concretely into a context of the bustle of all the overlapping practices in everyday life:

> I have worked many night shifts. Then it's not the great culinary expeditions I embark upon, it just isn't. . . . And the food gets marked by that

immediately, it does. Then it's pasta and minced meat and . . . what I have the energy to do.

The mothers have a whole vocabulary of categories expressing more or less legitimate deviances from what is considered 'real' homemade food and meals. The most used categories are 'popping over the lowest part of the fence', 'taking it on' and 'cheating'. Interestingly, one of these categories is explicitly used in the *Isabellas* media food representations as a heading for a regular feature in the magazine. This is the 'popping over the lowest part of the fence', and the feature pages represent food products and cooking procedures that are being framed as legitimate deviations from cooking properly from scratch – nearly homemade, that appears homemade.

So the lifestyle magazine media food genre seems to be domesticated among the mothers in three manners. First, the pictures and texts of the magazine are used as a resource to be appropriated in concrete food practices, knowledge and procedures in what and how to cook and serve homemade meals. Second, the magazine representations of meals and cooking become appropriated as an ambivalent category of proper food, both as a normative ideal and an unrealistic social expectation. Third, the food discourses of the magazine seem to also work as a resource for the mothers to legitimise practical adaptations and negotiations of such a normative ideal of cooking from scratch.

Similarities in mothering across media food genres

The above two subsections showed the differences in the domestication of the media food representations among the participants from the health campaign study and the lifestyle magazine study. The constellation of categories of food and cooking in the mothering seems to be somewhat related to the difference in genre between the two types of media food. However, analysis of the two data materials also shows distinct similarities across the two media contexts and social groups. In the following, I will highlight four of such similarities – mothering as loving, mothering as protecting, mothering as identification, and inappropriate mothering[9] – and exemplify them from both media food contexts. The concrete quotes do not necessarily all address or include media food representations directly, due to the non-media-centric practice theoretical approach to food and media use as intersecting everyday life activities.

Mothering as loving

This is a way of performing mothering which is described quite extensively in the research on food in everyday life. Here food is used for building and maintaining family relationships (Warin et al., 2008), and cooking labour turns food into homemade gifts of love (Moisio et al., 2004, p. 374; Ristovski-Slijepcevic et al., 2010, p. 476).

In both studies, the women with children cook meals they know the children like and serve other food items or meals for the children than the rest of the family is having. One of the Pakistani Danish mothers, Shabana, expresses the engagement of serving what the children like quite clearly: 'If we have friends of the children over, off course we make a bit more out of it. If the children like pizza, they get pizza, if they like schawarma, they get schawarma.' One of the magazine readers, Ellen, puts it like this: 'And often the children go yum-yum, and you can see they shove it in, and that is a kind of satisfaction.'

What is served for the children is clearly a normative issue with ambivalences and negotiations between food as love and relationship, and food as contested category. One of the other Pakistani Danish women gives her children chicken nuggets when the parents want to eat something the children don't like, 'but they cannot have it every day' (Solejma). The addition that the children cannot have it every day indicates the normative unease at serving something different and otherwise categorised as unhealthy food for the children (see Anving and Thorsted, 2010).

Another way of seeing the relation between mothering as loving and normative issues in food is when mothering through food is not regarded as sufficiently loving. Here, Thea explains about an episode relating to a Christmas meal in her parents' home, where the gift of love aspect of food is particularly important due to the highly ritualised character of meals:

> We were over there for x-mas, and she [Thea's mother] had, yes she had bought ready-made rice porridge in one of those plastic bags, right, to make the risalamande,[10] and it was really like this, when I opened the fridge, they were just lying there, and I was just like, okay we are going to have a good x-mas, I am not going to make a scene, and then I just went out to Emil and said, do you know what's in the fridge (laughs). . . . She obviously doesn't bother to make the porridge . . . it didn't taste too good either, that was really a bit of a miss, not stirred with love!

The lack of loving in this example is due to the lack of time sacrificed by the mother to prepare this traditional Christmas dessert. This way the normative expectation of a dish cooked from scratch by mother is not performed (see also Moisio et al., 2004, pp. 370–371; Molander, 2011, p. 85).

Mothering as protecting

In the existing literature on food, families and gender, to enact protection of children through provisioning of food and cooking is often seen as an aspect of managing or regulating children's eating (Coveney, 2000). Mothers see themselves as 'default family food managers' (Blake et al., 2009, p. 6), and claim that otherwise their families would eat 'junk' (Beagan et al., 2008, pp. 661–662). The mothering as protecting is closely related to enhancing healthier eating with the children.

Solejma, one of the Pakistani Danish women, explains in details about concrete procedures of cooking which are employed in order to protect her children from too much fat in the dinner meals:

> Ehm, not so much oil, I don't do that either. You know regarding the procedure, I think a lot about how to do it . . . ehm, for example our Pakistani rice, the way I make it for my children, I boil them first. And then I mix the other things into it. But if I do it the Pakistani way, then it's over-greasy.[11]

In a similar vein, Dorte, one of the lifestyle magazine readers, explains about protecting her daughter from eating too much sugar and artificial ingredients:

> It's always been like this with me that if I just looked at a cream puff, then I've put on weight, and unfortunately Paula has inherited this, so I'm very conscious about . . . what and how she eats, and we have a very conscious candy-policy in our family. . . . If I'm having something, it has to be a pure piece of chocolate or cake and none of all that synthetic stuff . . . argh, I can't have it. And then it's, you don't know what it does to the children, both in relation to sugar and chemicals and such, right?'

The existing literature also report upon mothering through food as protection by adapting and controlling children's eating habits, both in relation to healthier food discourses (Ristovski-Slijepcevic et al., 2010) and in relation to normative ideals about cooking meals from scratch (Molander, 2011, p. 85).

Mothering as identifying

The existing literature connects mothering processes closely with female identification, intersecting with other elements of identification such as class and ethnicity (Skeggs, 1997; West and Zimmerman, 1987). Food is part of these identification processes in mothering (e.g. Warin et al., 2008), but at the same time, food and eating itself is considered a major element of social and personal identification in the social and cultural scientific literature on food (e.g. Ashley et al., 2004; Bourdieu, 1984; Lupton, 1996; Murcott, 1983; Warde, 1997).

There are two main ways in which performing mothering through food is enacted as identifying among the women in the two Danish research projects. The first is when mothering is connected with belonging. Belonging in performing mothering through food is enacted in relation with the nuclear family of the women, the social group or network, and the cultural categories and practices they identify with. The two first empirical sections of this chapter show several examples of this. For example, where two of the Pakistani women talk about lasagne, and how 'we use our own spices'. This is an example of how belonging weaves together social group and cultural category of food. Another example is where belonging weaves together family biography and cultural category of food. Pia, one of the magazine readers, explains how making homemade pastry was a way of providing food that she learned in her childhood and has taught her own daughters.

The other main way in which mothering through food is enacted as identifying is when mothering is connected with social status and distinction (Vincent and Ball, 2007). An example among the Pakistani Danes is when Solejma explains about where she gets her understandings about healthier food from, and she positions 'sisters-in-law'[12] as a social group less knowledgeable:

> And if we search on the internet, you know, we get to know from many different places, also from the media, yes, . . . ehm, in contrast to us, our sisters-in-law and the others who are not so much into this, they have to learn – they have to be taught a bit.

Solejma understands her own mothering as somebody who is capable of knowing how to cook healthier food for her family, and she distinguishes her own mothering from a social group whom she positions as not so capable at mothering properly through food nor capable of using media for help. There are also examples where the women position themselves in a more ironic manner in a particular social status group. Here, it is the magazine reader, Ellen:

> Right, yes, I thought it was really funny when coming down to Anders' school, and you knew that you had got this vegetable box from Årstiderne,[13] very nice with recipes, and you had got those turnips, and then when we put all the food on a common table at the school, you could see that there where at least three of us who had made the same salad with vinaigrette and the turnips from the Årstidernes box, and it's just . . . oh no (laughs) . . . it just becomes so politically correct and predictable.

Ellen understands herself as somebody who can be identified as a middleclass mother with tendencies to buy organic food and wanting to display this way of doing mother to other mothers. A British study showed that women were constructing their families and what family meant to them through displaying to others what they understood as proper family meals (James and Curtis, 2010).

Inappropriate mothering

Normative questioning of food practices of the mothers is, not surprisingly, an issue in both studies. The negotiations and micro regulations of the practical moralities in mothering through food is recognizable from other qualitative studies of food and gender in the international literature (e.g. Bugge and Almås, 2006; Coveney, 2000; Moisio et al., 2004; Ristovski-Slijepcevic et al., 2010).

In the following example from a family interview with two of the Pakistani Danes, it is the mothering through food of one of their sisters that is being constructed as inappropriate. The inappropriate performance here is that the taste of the meal is given a lower priority vis-à-vis an ability to signal that the meal is healthier.

MARIA: 'My sister, do you know what she does? She only uses two teaspoons [of oil], and then when the onions have coloured, she takes the oil out and throws it away. And then she finishes the dish, that's why her food tastes too bad. . . .'
SADA: 'That's not good. That definitely doesn't taste nice.'
MARIA: 'No, it doesn't taste good, but then she feels she has done a good deed, right? . . . NOW we're eating healthy!'

Thus, to practice 'good' mothering through food can be quite a difficult task, with competing and intersecting normative engagements to live up to. Furthermore, there may not always be consensus in the family or social network on what constitutes proper mothering through food. In the following example from the lifestyle magazine project, the network group of women in their 40s enacted a normative disagreement on using fish fingers for family meals:

SONJA: 'When I was a child, when we had fish it was these frozen fish fingers . . .'
KAREN [INTERRUPTS]: 'Yes, THAT is disgusting [Sonja laughs] . . . that is bad cooking.'
DORTE: 'Yes, it is.'
SONJA: 'That is simply YUKKY.'
BIRGIT: 'Yes, my kids love them . . .'
DORTE [INTERRUPTS]: 'No, nobody likes fish fingers.'
BIRGIT: 'Yes, I like them.'

After having attempted to legitimate her use of fish fingers, Birgit only manages to get back into the discussions in the group after a while, and by way of agreeing to how some other cooking process (here deep frying) belongs to the category of 'bad cooking' that she does not perform.

Conclusive discussion

It is not news that the connections between food and motherhood are fraught with normative expectations. But contemporary women in Denmark perform mothering in everyday lives where different normative media food discourses are ever more available and ever less escapable, due to the media development and the mediatisation of culture (Hepp, 2012). This chapter deals with two studies of Danish women (Pakistani Danes and women aged 25–50) and the relations of their mothering to different food contesting media food genres. The two media food genres are public health campaigns and lifestyle magazines.

The patterns in the two empirical studies seem to suggest that the specific media food framing seems to some degree to be domesticated adapted, negotiated and made their own – as a symbolic resource for forming the specific social space for mothering through food practicing in each of the two social groups of participants. Hence, among the Pakistani Danish mothers, doing proper feeding of the family gets partially framed through the same types of categories for unhealthy and healthy food as the health campaign, although doing healthier food gets negotiated with e.g. proper Pakistani food. Among the magazine readers, doing proper feeding of the family gets partially framed through the negotiable category of cooking from scratch, which is negotiable in the specific lifestyle magazine as well. In both cases, one of the ways in which the hybrid and negotiated types of mothering through food could be interpreted is as possibly heterotopian (Foucault, 1984; Meininger, 2013) performances. One example of such mothering including 'othering' could be the magazine readers' blending of cooking from scratch normality with the socially out of place 'cheating' in using semi-produced and ready-made foods. Another example could be the Pakistani Danish mothers' mixing and shifting between Pakistani meals as good food normality and out of place healthier food.

On the other hand, there are important similarities in ways of performing mothering through food across the two media food genres and the two social groups. Mothering through food is about loving, protecting and socio-cultural identification, and mothering through food is negotiated and re-framed in relation to normatively inappropriate cooking and feeding activities. The question is whether it is possible to understand some of these similarities in mothering through food as placed in relation to more hegemonic societal discourses (Foucault, 1978). Such a question is impossible to answer on the basis of only two qualitative case-studies. But the parallels with empirical findings in the existing research literature on mothering as loving, protecting and gender identification suggest that this could be argued. However, from my practice theoretical perspective, such discourses will probably most often become adapted, re-invented slightly, negotiated etc. in the processes of domestication, because domestication of media food discourses take place embedded in a family everyday life (Hartmann, 2013), intersected by many different practices and normative engagements.

My understanding of one of the consequences of mediatisation to be a blending of media use into all other everyday practices (Couldry, 2004, p. 125; Hartmann, 2009, p. 422) tended to inform the analytical design of both empirical studies. Thus, the analyses of mothering through food relations with media food genre are based upon a non-media-centric understanding of media in everyday life (Pink and Mackley, 2013, p. 681). Neither of the two studies was designed as reception research or food media use research with the campaign material or the magazine at the centre of the data-production. Rather, the media food genre in question worked as a representational sounding board in both studies.

This does not of course mean that the differences in media food genre are not important. The Pakistani mothers draw upon wording, phrasing and imagery of healthier food, parallel to the representation of such food in public campaign and dietary advice material. They also negotiate the position of themselves as somewhat responsible for changing eating habits in the families, a responsibility which is often part of the framing of healthier food campaigning (Vallgårda, 2007). The female magazine readers draw upon wording, phrasing and imagery on proper food and good cooking, parallel to the representations of food and cooking from the lifestyle magazine. The magazine is also domesticated among the women along forms recognisable for women's magazines in general, such as relaxation and practical knowledge (Hermes, 1995).

On the other hand, the media food genres themselves are perhaps becoming blurred at the edges, in line with the apparent media platform convergences (Jensen, 2010). Public campaigning on food nowadays apparently tend to draw upon several media food genres, mixing e.g. entertainment with information. And lifestyle magazines seem to be blending in food issues from across the range of media food genres. Furthermore, parts of current empirical media research argues that it is difficult to presume citizen attention to any kind of public issues in media, and researchers should rather talk about some kind of looser 'public connection' with different issues through media (Couldry et al., 2010). In that light, none of the two types of media food genres necessarily attract more privileged domestication than the other. But more research is needed in this area. It is argued in current communication and media research that the widespread framing research that deals with framing of content in media representations is in need of more socio-culturally theorised studies of audience frames (Vliegenthart and van Zoonen, 2011). It could be argued that investigating mothering through food in relation with the representational sounding board of the cross-genre media food 'soup' in everyday life could be seen as an example of such a way of studying audience frames, although a non-media-centric one.

The relationship between mothering through food and media food genre domestication in everyday life is perhaps not so straightforward. With very good reason, some media researchers may tend to assume that particular genres and media use activities can work as more influential appropriations than others. On the other hand, everyday practitioners tend to enact their food and mothering across these categories in what they do and what they draw upon as symbolic resources in their domestication of media food. They do so in order to juggle their bodily and discursive relations with a multiplicity of media food framings as well as a multiplicity of overlapping different practices, identifications and normative expectations. In the light of the current media development with cross-genre uses, mobile media platforms and extensive media saturation, the challenges in everyday juggling of mothering through food are not likely to shrink.

Notes

1 The research project was financed by the Danish Social Scientific Research Council (FSE), 2008–10, and carried out together with Iben Jensen, Ålborg University.
2 The research project received support from Roskilde University, 2006–8.
3 Neither of the two studies were originally focusing on mothering and media use. The importance and varieties of mothering through food were empirical outcomes of abductive (Blaikie, 2007) qualitative research in both studies.

4 What is 'healthy' and 'unhealthy' food is what is constructed as such by the participants. These understandings, however, do come rather close to the official Danish nutritional advice (Andersson and Bryngelsson, 2007, pp. 36–38).
5 Already here is a cultural hybridised format where the generalised Italian dish becomes Danish.
6 All names are pseudonyms.
7 Stewed strawberry is a traditional Danish dessert where strawberries are cooked together with sugar to a kind of porridge and served hot or cold with milk or cream.
8 For a theoretical discussion of the concept 'do-ability' in a practice theoretical approach, see Halkier (2010), p. 36.
9 Readers who are interested in a more detailed presentation of the different types of mothering, please see Halkier (2013).
10 Risalamande is the most traditional Christmas dessert in Denmark, made from cold rice porridge mixed with blanched chopped almonds, whipped cream and vanilla and served with warm cherry sauce.
11 This is called biriani rice, where the rice is steamed, and the oil-roasted vegetables and spices are made separately and mixed into the rice just before serving. Whereas pilao rice, the rice are roasted together with the vegetables and spices, requiring more oil in order for the rice not to stick to the pot.
12 Sisters-in-law to women born in Denmark are often born in Pakistan.
13 Årstiderne is an organic box delivery programme in Denmark. Årstiderne means 'the seasons'.

References

Andersson, A. and S. Bryngelsson, 2007. Towards a healthy diet: From nutrition recommendations to dietary advice, *Food & Nutrition Research*, 51(1), 31–40.

Anving, T. and S. Thorsted, 2010. Feeding ideals and the work of feeding in Swedish families: Interactions between mothers and children around the dinner table, *Food, Culture & Society*, 13, 29–45.

Ashley, B., J. Hollows, S. Jones and B. Taylor, 2004. *Food and Cultural Studies*. London, Routledge.

Atkinson, R., 1998. *The Life Story Interview*. London, Sage.

Barbour, R., 2007. *Doing Focus Groups*. London, Sage.

Beagan, B., G.E. Chapman, A. D'Sylvia and B.R. Bassett, 2008. 'It's just easier for me to do it': Rationalizing the family division of foodwork, *Sociology*, 42, 653–671.

Bennet, W.L. and S. Iyengar, 2008. A new era of minimal effects?, *Journal of Communication* (58)1, pp. 707–731.

Bird, S.E., 2010. From fan practice to mediated moments: The value of practice theory in the understanding of media audiences, in B. Bräucher and J. Postill (eds.): *Theorising Media and Practice*. New York, Berghahn, pp. 85–104.

Bjur, J., K.C. Schrøder, U. Hasebrink, C. Courtois, H. Adoni and H. Nossek, 2013. Cross-media use. Unfolding complexities in contemporary audiencehood, in N. Carpentier, K.C. Schrøder and L. Hallett (eds.): *Audience Transformations: Late Modernity's Shifting Audience Positions*. London, Routledge, pp. 15–29.

Blaikie, N., 2007. *Approaches to Social Enquiry* (2. Ed.). Cambridge, Polity.

Blake, C.E., C.M. Devine, E. Wethinton, M. Jastran, T.J. Farrell and C.A. Bisogni, 2009. Employed parents' satisfaction with food-choice coping strategies: Influence of gender and structure, *Appetite*, 52, 711–719.

Bourdieu, P., 1984. *Distinction: A Social Critique of the Judgement of Taste*. London, Routledge and Kegan Paul.

Bourdieu, P., 1990. *The Logic of Practice*. Cambridge, Polity.

Bugge, A.B. and R. Almås, 2006. Domestic dinner: Representations and practices of a proper meal among suburban mothers, *Journal of Consumer Culture*, 6, 203–228.

Butler, J., 1990. *Gender Trouble: Feminism and the Subversion of Identity*. New York, Routledge.

Caldwell, M.L., 2004. Domesticating the French fry: McDonald's and consumerism in Moscow, *Journal of Consumer Culture*, 4, 5–26.

Clarke, A.J., 2004. Maternity and materiality: Becoming a mother in consumer culture, in J.S. Taylor, L. Layne and D.F. Wosniak (eds.): *Consuming Motherhood*. London, Rutgers University Press, pp. 55–71.

Couldry, N., 2004. Theorising media as practice, *Social Semiotics*, 14, 115–132.

Couldry, N., S. Livingstone and T. Markham, 2010. *Media Consumption and Public Engagement: Beyond the Presumption of Attention*. Basingstoke, Palgrave Macmillan.

Coveney, J., 2000. *Food, Morals and Meaning: The Pleasure and Anxiety of Eating*. London, Routledge.

DeVault, M., 1994. *Feeding the Family: The Social Organization of Caring as Gendered Work*. Chicago, The University of Chicago Press.

Eden, S., 2009. Food labels as boundary objects: How consumers make sense of organic and functional foods, *Public Understanding of Science*, 18, 1–16.

Floyd, J. and L. Forster, 2003. *The Recipe Reader: Narratives, Contexts, Traditions*. Aldershot, Ashgate.

Foucault, M., 1978. *The History of Sexuality. Vol. 1*. Harmondsworth, Penguin.

Foucault, M., 1984. Of other spaces, heterotopias. *Architecture, Movement, Continuité*, 5, 46–49.

Frey, J.H. and A. Fontana, 1993. The group interview in social research, in D.L. Morgan (ed.): *Successful Focus Groups*. London, Sage, pp. 24–34.

Giddens, A., 1984. *The Constitution of Society*. Cambridge, Polity.

Halkier, B., 2010. *Consumption Challenged: Food in Medialised Everyday Lives*. Farnham, Ashgate.

Halkier, B., 2013. Contesting food – contesting motherhood?, in S. O'Donohoe, M. Hogg, P. Maclaran, L. Martens and L. Stevens (eds.): *Motherhoods, Markets and Consumption. The Making of Mothers in Contemporary Western Cultures*. London, Routledge, pp. 89–134.

Hammersley, M. and P. Atkinson, 1995. *Ethnography: Principles in Practice*. London, Routledge.

Hartmann, M., 2009. The changing urban landscapes of media consumption and production, *European Journal of Communication*, 24, 421–436.

Hartmann, M., 2013. From domestication to mediated mobilism, *Mobile Media & Communication*, 1, 42–49.

Heisley, D.D. and S.J. Levy, 1991. Autodriving: Photoelicitation technique. *Journal of Consumer Research*, 18, 257–272.

Hepp, A., 2012. *Cultures of Mediatization*, Cambridge, Polity Press.

Hermes, J., 1995. *Reading Women's Magazines: An Analysis of Everyday Media Use*. Cambridge, Polity Press.

Hollows, J., 2003. Feeling like a domestic goddess: Postfeminism and cooking, *European Journal of Cultural Studies*, 6, 179–202.

Hollows, J., 2007. The feminist and the cook. Julia Child, Betty Friedan and domestic femininity, in E. Casey and L. Martens (eds.): *Gender and Consumption: Domestic Cultures and the Commercialisation of Everyday Life*. Aldershot, Ashgate, pp. 33–38.

Holm, L., 1996. Identity and dietary change, *Scandinavian Journal of Nutrition*, 40, supplement 31, S95–S98.

Holm, L., 2013. Sociology of food consumption, in A. Murcott, W. Belasco, and P. Jackson (eds.): *The Handbook of Food Research*. London and New York: Bloomsbury Academic, pp. 324–337.

Holstein, J.A. and J.F. Gubrium, 2003. Active interviewing, in J.F. Gubrium and J.A. Holstein (eds.): *Postmodern Interviewing*. London, Sage, pp. 67–80.

Hurdley, R., 2007. Focal points: Framing material culture and visual data, *Qualitative Research*, 7, 355–374.

Jabs, J. and C.M. Devine, 2006. Time scarcity and food choices: An overview, *Appetite*, 47, 196–204.

James, A. and P. Curtis, 2010. Family displays and personal lives, *Sociology*, 44, 1163–1180.

Jensen, K.B., 2010. *Media Convergence: The Three Degrees of Network, Mass and Interpersonal Communication*. London, Routledge.

Johansson, B. and E. Ossiansson, 2012. Managing the everyday health puzzle in Swedish families with children, *Food and Foodways*, 20, 123–145.

Keller, M. and B. Halkier, 2014. Positioning consumption: A practice theoretical approach to contested consumption and media discourse, *Marketing Theory*, 14, 35–51.

Lupton, D., 1996. *Food, Body and the Self*. London, Sage.

Marshall, D., M. Hogg, T. Davis, T. Schneider and A. Petersen, 2013. Image of motherhood: Food advertising in 'Good Housekeeping' magazine 1950–2010, in S. O'Donohoe, M. Hogg, P. Maclaran, L. Martens and L. Stevens (eds.): *Motherhoods, Markets and Consumption: The Making of Mothers in Contemporary Western Cultures*. London, Routledge, pp. 116–128.

Martens, L. and S. Scott, 2005. "The unbearable lightness of cleaning": Representations of domestic practice and products in Good Housekeeping Magazine (UK), *Consumption, Markets & Culture*, 8, 379–401.

Meininger, H.P., 2013. Inclusion as heterotopia: Spaces of encounter between people with and without intellectual disability, *Journal of Inclusion*, 4, 24–41.

Mitchell, W. and E. Green, 2002. 'I don't know what I'd do without our mam': Motherhood, identity and support networks, *The Sociological Review*, 50, 1–22.

Moisio, R., E.J. Arnould and L.L. Price, 2004. Between mothers and markets: Constructing family identity through homemade food, *Journal of Consumer Culture*, 4, 361–384.

Molander, S., 2011. Food, love and meta-practices: A study of everyday dinner consumption among single mothers, in R.W. Belk, K. Grayson, A.M. Muniz and H.J. Schau (eds.): *Research in Consumer Behaviour*. Emerald Group Publishing, pp. 77–92.

Moloney, M. and S. Fenstermaker, 2002. Performance and accomplishment: Reconciling feminist conceptions of gender, in S. Fenstermaker and C. West (eds.): *Doing Gender, Doing Difference: Inequality, Power and Institutional Change*. New York, Routledge, pp. 189–204.

Montgomery, K.C. and J. Chester, 2009. Interactive food and beverage marketing: Targeting adolescents in the digital age, *Journal of Adolescent Health*, 45, s18–s29.

Murcott, A., 1983. 'It's a pleasure to cook for him: Food, mealtimes and gender in some South Wales households, in E. Garmanikow, D. Morgan, J. Purvis and D. Taylorson (eds.): *The Public and the Private*. London, Heinemann, pp. 78–90.

O'Sullivan, T., 2005. From television lifestyle to lifestyle television, in D. Bell and J. Hollows (eds.): *Ordinary Lifestyles. Popular Media, Consumption and Taste*. Maidenhead, Open University Press, pp. 21–34.

Peattie, K. and S. Peattie, 2003. Ready to fly solo? Reducing social marketing's dependence on commercial marketing theory, *Marketing Theory*, 3, 365–386.

Pink, S. and K.L. Mackley, 2013. Saturated and situated: Expanding the meaning of media in the routines of everyday life, *Media, Culture & Society*, 35, 677–691.

Povlsen, K.K., 2007. Smag, livsstil og madmagasiner [Taste, lifestyle and food magazines], *Mediekultur*, 23, 46–53.

Puchta, C. and J. Potter, 2005. *Focus Group Practice*. London, Sage.

Reckwitz, A., 2002. Toward a theory of social practices: A development in culturalist theorizing, *European Journal of Social Theory*, 5, 243–263.

Ristovski-Slijepcevic, S., G.E. Chapman and B.L. Beagan, 2010. Being a 'good mother': Dietary governmentality in the family food practices of three ethnocultural groups in Canada, *Health*, 14, 467–483.

Ritzer, G., 1993/2008. *The McDonaldization of Society 5*. Los Angeles, Pine Forge Press.

Schatzki, T., 2002. *The Site of the Social: A Philosophical Account of the Constitution of Social Life and Change*. Philadelphia, Pennsylvania State University Press.

Short, F., 2006. *Kitchen Secrets: The Meaning of Cooking in Everyday Life*. Oxford, Berg.

Shove, E., M. Pantzar and M. Watson, 2012. *The Dynamics of Social Practices: Everyday Life and How It Changes*. London, Sage.

Silverman, D., 2006. *Interpreting Qualitative Data*. London, Sage.

Silverstone, R., 1994. *Television and Everyday Life*. London, Routledge.

Skeggs, B., 1997. *Formations of Class & Gender*. London, Sage.

Spradley, J.P., 1979. *The Ethnographic Interview*. Fort Worth, Holt, Rinehart and Winston.

Thomas, G. (2011). *How To Do Your Case Study: A Guide for Students & Researchers*. London, Sage.

Thomson, R., L. Hadfield, M.J. Kehily and S. Sharpe, 2012. Acting up and acting out: Encountering children in a longitudinal study of mothering, *Qualitative Research*, 12, 186–201.

Vallgårda, S., 2007. Health inequalities: Political problematizations in Denmark and Sweden. *Critical Public Health*, 17, 45–56.

Vincent, C. and S.J. Ball, 2007. 'Making up' the middle-class child: Families, activities and class dispositions, *Sociology*, 41, 1061–1077.

Vliegenthart, R. and L. van Zoonen, 2011. Power to the frame: Bringing sociology back to frame analysis, *European Journal of Communication*, 26, 101–115.

Warde, A., 1997. *Consumption, Food and Taste*. London, Sage.

Warde, A., 2005. Consumption and theories of practice, *Journal of Consumer Culture*, 5, 131–153.

Warde, A., L. Martens and W. Olsen, 1999. Consumption and the problem of variety: Cultural omnivorousness, social distinction and dining out, *Sociology*, 33, 105–127.

Warin, M., K. Turner, V. Moore and M. Davies, 2008. Bodies, mothers and identities: Rethinking obesity and the BMI, *Sociology of Health & Illness*, 30, 97–111.

West, C. and D.H. Zimmerman, 1987. Doing gender, *Gender and Society*, 1, 125–151.

Windahl, S. and B. Signitzer, 2009. *Using Communication Theory: An Introduction to Planned Communication*. London, Sage.

10 Food across media

Popular food contents among children in Germany[1]

Susanne Eichner

'Life worlds are media worlds,' stated Baacke et al. in the early 1990s (Baacke, Sander and Vollbrecht, 1990), and in today's lifeworlds, ever more saturated with media, this is truer than ever. Children grow up using a variety of mass media and personalised media: they access entertainment or information at home and in school via the television, the computer, the smartphone, the games console and the tablet. In doing so, they come in contact with numerous representations of media food. Food and nutrition are an integral part of children's media environment and media products. Within this setting, children have been conceptualised as consumers (for example Buckingham, 2011), as patients (for example Diehl, 2005), or more generally as media users (for example Charlton and Neumann-Braun, 1990; Theunert et al., 1992; Vollbrecht, 2002). However, as Buckingham (2011, p. 5) states, within the field of media and food, research interest has tended to focus on the negative health effects of advertisements and marketing. This chapter leads in a different direction. It follows Wagner and Theunert (2007) in their approach of conceptualising children as recipients within increasingly convergent media environments, and begins with the basic question of where children meet food in the media they use, and how these food representations are staged and contextually embedded. The study presented here is based on a cross-media content analysis of the food-related media content that is most popular within the 6- to 12-year-old age group of children in Germany. Not only were the food-related media products identified and recorded, but also their 'mode of representation' and 'mode of appellation' were documented. The content analysis was further combined with a textual aesthetic analysis, elaborating on the 'immanent structure of a text' (Iser, 1980, p. 86) that provides the frame for the 'possibility space' of significance and negotiations for the audience. The research design thus aims to provide an empirical foundation on which to place the actual media 'doings' (Eichner, 2014) of children.

Food, media and children

Within the research area of communication studies and health communication, the relationship between food in media and children has been widely discussed especially in relation to television advertising and its potential negative effects on health (e.g. in Ashton, 2004; Cebulla-Jünger, 1995; Gamble and Cotunga, 1999; Hastings et al., 2003; Lewis and Hill, 1998; Vollbrecht, 1998). This research interest is important, considering the amount of time that young media audiences spend watching commercial television in most parts of the world. Yet other media and media content concerned with food and nutrition have only rarely been the focus of media and communication or of media studies. The concentration on advertising and its negative effects has resulted in a blind spot: while the relationship between food advertising and children has been well examined, though without clear results (Diehl, 2005), other areas of interaction between media food and children remain underdeveloped.

The study of health communication found its way into German academic institutions only with the new millennium (Rossmann and Hastall, 2013, p. 10). In 2003 the network Netzwerk Medien und Gesundheitskommunikation (Network Media and Health Communication) was established, in 2006 the book series *Medien + Gesundheit (Media + Health)* was launched, and in 2012 the Ad hoc group Gesundheitskommunikation (Health Communication) of the DGPuK (the German Society for Publicist and Communication Studies) was implemented. Particularly noteworthy in the context of this contribution are two strands of work: studies investigating the potentially positive effects of food and media (mainly television) for the population, and a second strand that focuses on the relevance and significance of media food for children in particular. Research from the first of these two strands is predominantly quantitative-based. For example Rössler et al. (2006) conducted a study on nutrition in German television. The study comprised a content analysis of eight German television channels, both public service and commercial, a telephone survey and a reception study. This was the first major study to map food representation in German television, and it demonstrated that food-related topics make up a high proportion of programming. In more than half the cases, food was a central object. Based on this study, Stephanie Lücke (2007) aimed to explore the relationship between food representation and the concepts, attitudes and behaviour of television viewers, in the tradition of cultivation theory (Gerbner, 1998). While both reports focus exclusively on television and on the impact of food representations on an adult audience, they provide a substantial overview of the German television landscape in relation to food. The study and its subsequent research thus provided a media food map of the German television landscape, bringing the significance of media food to consciousness. Also the education–entertainment approach,[2] as taken for example by Arendt (2010, 2013) and Lampert (2007), unfolds within this strand of research.

The second strand of ongoing research can be traced in the work of the IZI (the International Central Institute for Youth and Educational Television),[3] where a qualitative and audience-oriented approach is employed. Exemplary here is the explorative study by Götz, Holler and Unterstell (2010), which stressed that narratively embedded health messages were memorised in greater detail by children than messages within a non-narrative context. This supports earlier findings such as by Jensen (1999), who argued that news became meaningful for audiences when linked to the recipients' personal narrative.

Approaches within the field of health communication are often strongly effect-driven. While many studies focus on the negative effects of food representation on health, more recent studies in Germany focus on the positive potential of media as a device for health education. Where and how, so goes the underlying question, can people best be reached and impacted by health messages in order to prevent obesity and other food-related diseases? Additional research often centres on the media use of adults or adolescents. This results in an academic void concerning questions of media use of children in relation to nutrition topics, with the very basic issues remaining unstated: what kind of media food do children come in contact with in the course of everyday life media consumption? How is the food-related content media-specific? How is this content staged and moderated? Where and how are food-related messages moderated? What significance do the food representations hold for children? What is their role in the processes of reception and appropriation, the 'complex, subjectively varying process of integrating media in the context of everyday life and experience' (Theunert, 2005, p. 115)? The current study cannot answer all of these questions. But it will add one component to a better understanding of the field.

Research design and methodology

Media food is not isolated from other representations or issues. It is embedded in a diversity of formats, genres, dramaturgies, narrations and statements. Moreover, while many studies focus on the mechanisms of one specific, isolated medium, such an approach does not reflect the actual media user-patterns of children – the whole processes of media use from selection and reception to appropriation. As Wagner and Theunert (2007) pointed out, children grow up in increasingly convergent media environments and do not use different media in isolation. Rather, media use in everyday life takes place within a media ensemble and has its specific pattern with regard to age, gender or educational background of children, which again reflects back on the 'meaningful media experiences' (Wagner and Theunert, 2007, p. 16, p. 20). Furthermore, convergence, in a context of practise theory (Reckwitz, 2002), emphasises that cultural practices – such as media reception – are positioned within an array of medial and non-medial activities which 'coexist and converge' (Hengst, 2014, p. 19). At the same time, economic practices interconnect with our lifeworlds. Franchise strategies have for a long time catered for a spill-over of media products, characters and objects into the 'real' world. Hollywood's strategy of merchandising, spin-off production and cross-media integration, with its paradigmatic and groundbreaking example of *Star Wars,* has long been adopted by television and other media. In 2014, for instance, the green frog monster Om Nom, known from the bestselling popular mobile puzzle-game *Cut the Rope* (Zeptolab and Chillingo, 2010), can be purchased in the shape of an apple box at McDonalds.[4] And *Connie* (Carlsen Verlag), a popular German children's book series with the protagonist Connie, returns as game-based learning app on the family smartphone or tablet. Against the background of converging media environments and converging media practices, the survey was designed as a cross-media study, covering the most popular media for children.

The first step of the study was to identify the existing food- and nutrition-related cross-media content with high audience shares or high sales figures among 6- to 12-year-old children. The premise thereby was not to focus automatically on media products and content aimed at children, but to identify media products and contents actually watched, read or played by children, whether or not they were the intended audience. These are not necessarily the programmes that are produced for children. As television audience ratings show, children tend to migrate to other programme offerings during their pre-teen phase: correspondingly, formats of the 'normal' adult programme scheme are among the highest-rated programmes for pre-teens: *Deutschland sucht den Superstar* (RTL), *Germany's Next Top Model* (pro7), *Ich bin ein Star – holt mich hier raus* (RTL), *The Voice Kids* (Sat1), and big sports events, such as important champions-league games, are favoured not only by an adult audience but also by children (Hofmann, 2012, p. 29; KIM-Studie, 2012, p. 19; AFG/GfK). This trend can be witnessed across media: the most popular websites among pre-teens are *YouTube* and *Facebook,* and the most popular video games are also targeted towards an older audience (KIM-Studie, 2012, p. 38, p. 52).

Consequently, the sample of the content analysis was selected by means of highest range within the target group of 6- to 12-year-old children, as well as by means of the commitment shown by children to media products. These two premises were operated and identified on the basis of detailed television ratings (AFG/GfK); the KIM-Studie 2012 very extensive and long-lasting study of children's media use and their preferences in Germany;[5] two consumer surveys (KidsVA, 2012; Trend Tracking Kids®, 2012) that provide information on sales figures and on children's preferences concerning consumer products and purchase media; and three big online branch observers *(www.mediabiz.de; www.vgchartz.com; www.buchreport.de)*. The most relevant media identified were television, video games, internet, books, comics, DVDs, children's journals and flyer/brochures.[6]

In a second step, the relevant programmes and products were selected. With regard to television, the most popular and highest-rated channels within the target group were chosen on the basis of ratings within the relevant age group. These are the two commercial full programme channels RTL and pro7, the public service children's channel KiKa, and two commercial children's channels SuperRTL and Nickelodeon. All five channels were recorded in a randomly selected sample week (9–15 December 2013) during the day curve peak between 18:30 and 20:45 (Feierabend and Klingler, 2013, p. 194) and additionally during the weekend peak (ibid.) in the morning. All other media samples were added according to a combination of sales figures and audience surveys. With regard to the study's aim and possibilities, the respective three most popular products within each medium were identified and added to the sample.[7] The final total sample consisted of more than 200 hours of television, four major video games, 11 internet sites, 3,860 book pages, 1,401 pages of children's journals, 471 minutes of DVD and 338 pages of flyers.

The cross-media design, however, created methodological problems. For example, for television detailed audience information concerning ratings and viewing time, age and gender were available, whereas for other media either only sales figures or a combination of sales figures and audience surveys could be consulted. This not only results in a bias that might distort the actual media use of media other than television by children but also stages a quantitative problem: television programming makes up by far the biggest share of the sample. While this matches to some extent the media use of children in Germany who prefer television above all other media and most other leisure-time activities (KIM-Studie, 2012), it does not reflect the precise amount of time that children spend with the different media. Yet in relation to the purpose of the overall study – to map out an initial draft of the media landscape of food-related *content* – the methodological problems were considered as permissible. Ultimately, the aim of this stage of the study was to provide an overview of cross-media media food: the actual use is to be covered by a subsequent audience study.

Building on the quantitative data, qualitative aspects were built into the operationalization of the material to ascertain how the food- and nutrition-related content is presented to the audience. Thereby, not only the sheer quantitative number of media food contacts across media was captured, but also their structures of appeal, their specific generic context and their audio-visual strategy were mapped. This depicts the landscape of media food for children in greater depth and sheds light on the various modalities of existing media food across media. The following five categories served this purpose. Firstly, the 'kind of food' and their representational mode were recorded. A fruit, for example, could be represented only verbally or only visually, or it could be referred to only text-based, or various different modes of representation could be accumulated in an audio-visual representation combined with text. The 'genre or type' of a media product was then identified, relating the mode of food representation to an overall structure of generic appellation of the media product. Following this step, the 'mode of appellation' within each unit of analysis was classified. This refers to the character of the address to the audience, whether this is in the form of a narrative structure, an informative–factual or informative–entertaining structure, a ludic structure, an interactive structure or an advertising structure. The modes of appellation are thus independent of the overall genre or type of 'programme,' since it was assumed that a fictional television programme could contain elements addressing its audience with a ludic or interactive structure; an online game, on the other hand, might include elements involving narrative structures. Next, it was questioned whether the relevant unit comprises food representations that are 'dominant,' in the sense that they are the focus of attention, or whether they are just one aspect among others, for example, part of the setting. Finally, given the overall aim of the study, any 'intended educational claim' of food representations was identified. Some of the central findings are presented in the following section.

Content analysis: overview

The whole sample consisted of 785 programme units, of which 206 (26 per cent) featured a food or nutrition reference. Within these programme units 1,156 units of analysis[8] were identified, of which 546 were within the main data consisting of media without advertisements and 'serious media.' Television was also the biggest contributor of food representations, as is consistent with its key role in the media ensemble of children in Germany; Feierabend and Klingler see this as its 'outstanding role' (2013, p. 190). Nearly two-thirds of all units of analysis (68.9 per cent) could be located within television. Books came in second place with 12.1 per cent. Comics and journals together with the internet also display a noteworthy amount of food representations (7.3 and 6.2 per cent). Video games, DVDs and flyers play only a minor role in food representations, but because of the small sample of only three titles, the quantitative results here are only of limited informative value.

Within most media the highest proportion of represented food were sweets, followed by beverages (often sweet drinks), fruits and vegetables. Clearly, this does not support the recommendation of the DEG, the German Society for Nutrition. The results are remarkable, because the underlying main data were cleared of all advertisements. Sweets are the most frequently represented kind of food in television, on the internet, in flyers and also in books. In comics, they are in second place after prepared dishes. The only exception is DVDs, where sweets are number four after beverages, general nutrition and meat. Very often beverages are the second most frequently presented nutrition (television, internet, books). The variety of represented food is thus media-specific: television and the internet display the greatest variety of food (14 distinguishable categories), followed by books (11 categories) and comics and journals (8 categories). DVDs and flyers show only a limited variety of food representations. The strong focus on sweets and beverages might – beyond nutritional recommendations – be due to the special interest children show in sweets. They stage a connection point to their own interests and preferences.

Table 10.1 Units of analysis with food references

Units of analysis with food reference (Main data)	*Frequency (in UoA)*	*Frequency (in %)*
Television	376	68.9
Books	66	12.1
Journals/comics	40	7.3
Internet	34	6.2
DVDs	15	2.7
Flyers/brochures	13	2.4
Video games	2	0.4
Total amount	**546**	**100.0**

Important for the study was the context and the mode of representing food, conceptualised with the categories 'genre' and 'mode of appellation.' The methodological challenge was to formulate media-*unspecific* categories. The mode of appellation catered for this aim and was designed as a media-unspecific category, while genre tends to be highly media-specific. Genre thus served to give a general overview of where to locate food representations, while the mode of appellation describes the sample qualitatively. Children's media in Germany – and here especially television – are dominated by narrative structures often displayed by fictional formats. According to Krüger (2009), all children's channels have a very high proportion of fiction: leading here is the commercial children's channel *SuperRTL,* where more than 90 per cent is fiction (Krüger, 2009, p. 419). Narrative dominance can also be detected in popular books, comics and DVDs. Accordingly, animation displays the highest proportion of food representation within the sample (43 per cent), followed by informal education TV programmes such as the popular children's format *Wissen macht Ah!* (KiKa) or the infotainment format *Galileo* (pro7). These and similar formats have become very popular in Germany's contemporary children's television, known by the term *Wissenssendungen.* The third biggest share of food representations were found in books, within the genre of children's and teen novels.

Not surprisingly, the structures of appellation largely match the identified genres: more than two-thirds of the units of analysis display a narrative structure (77 per cent), followed by informative–entertaining structures (17 per cent). Other modes of appellation are only rarely employed. When food is represented in children's media, its embellishing function is usually one aspect among others and thus it is considered in this study as being not dominant. Yet in one-third of the data, food representation does in fact become the dominant focus of concern. The genres and types of media most frequently displaying a dominant food representation are animation, informal educational TV programmes and the teen novel. Only a very small percentage of food-related presentations claim to be educational (8 per cent).

Summing up the outcomes of the content analysis, when children between the ages of 6 and 12 use media in Germany, they most likely encounter representations of sweets, beverages, fruits and vegetables. Very probably, these representations are embedded in a TV animation, an informal educational programme on TV, or a children's or teen novel. Contributions on the internet were only marginal within this study. This is due to the method of measurement: while the internet indeed provides an enormous number of food-related offers for children, here only the most popular websites were added to the data set. The most frequent mode of appellation in connection with food representations is either narrative or informative–entertaining, usually embedded in a factual format. One-third of these representations feature a dominant focus on food. Animation formats like *The Penguins of Madagascar* (Nickelodeon), *Shaun the Sheep* (KiKa), or *Curious George* (KiKa); factual TV shows such as PUR+ (ZDF) or *Galileo* (pro7); or teen novels such as Rick Riordan's *Percy Jackson* and *Heroes of the Olympus* series, are typical examples of 'contact points' between children and media food.

Food representation and practical significance

Two formats – one television format and one book format – serve as examples for a further qualitative text analysis that complements the quantitative content analysis and specifies the potential space of meaning-making provided by the immanent textual structures, their specific aesthetic patterns of appeal (Mikos, 2001, pp. 177–178). In the tradition of reception aesthetics and neo-formalistic film theory, this is based in a conception of text–recipient interaction that conceptualises the text as a 'semiotic resource' (Fiske, 2009, p. 64) and a 'semiotic space' (Fiske, 2009, p. 85) that guides and cues reception processes. In this sense, media texts are considered to have an 'affordance character' (Iser, 1970; Mikos, 2001, p. 179) that pre-structures the reception. Reception in this connection is considered neither to be determined by the text, nor arbitrarily subjective, but the 'medial representation as an aesthetic structure has to be considered as a functional element of reception' (Mikos, 2001, p. 286).

Media texts consist of narrations, representations, simulations and information – in short, of the whole array of symbolic material that is available. People respond to media texts cognitively, emotionally, habitually and ritually, and integrate them practically and habitually in their everyday life (Eichner, 2014). In the phase of appropriation, the content is assimilated to one's own life situation: this represents the phase of practical sense (Charlton and Neumann, 1988). Food representation can take on specific functions within the processes of media reception and appropriation, within the whole course of media 'doings.' In general, media can offer orientation and identification (Paus-Haase and Hasebrink, 2001), they structure daily life, they offer information and satisfy cognitive and emotional needs, and they serve as demarcation from parents or adults in general (Vogelgesang, 1991). During the phase of adolescence they are important for distinguishing processes, as well as specifying exclusion and inclusion group processes (Niesto, 2009). Finally, media enable reflexivity by providing multiple informations, multiple lifestyles and multiple life scripts (Wegener, 2008, pp. 36–37). Food representation relates to most people's everyday life. However, the specific function depends on two factors: firstly, the concrete aesthetically and dramaturgically staged representation within a specific context such as genre; and secondly, the specific tactics and practices that recipients employ within the context of their specific lifeworlds and life phases. Media and texts pre-structure not only the act of reception, but the whole process of reception, appropriation and meaning making, conceptualised as media 'doing' (Eichner, 2014). The following textual analysis will carve out the space of possibilities as described.

Orientation, information and gratifications in *Shaun the Sheep*

This stop-motion animation series by Nick Park has been broadcast since 2007 on the German PBS-operated children's channel KiKa. The series appeals to adults and children alike because of its multi-layered narration and humour. It is screened occasionally as a two-part series in the time slot from 18:05 until 18:15. Young girls, aged 6 to 9 years in particular, like the channel and provide a high audience share (AGF/GfK). In the episode *Shaun the Sheep – Shape Up with Shaun* (KiKa, 13 December 2013), Shaun and his friends develop a workout for the largest sheep of the flock, Shirley, since she can no longer squeeze through the barn door. After she has been successfully transformed into a thin athlete, Shirley is accidentally thrown into a passing truck loaded with cookies and doughnuts. When the flock rescues her, she has already eaten half the cartload and is back to her old size.

Characteristically, the series communicates its story audio-visually, without employing dialogue. The regular characters consist of the inventive sheep Shaun, the flock, the sheepdog Bitzer and the farmer. The theme of this episode is the relationship between sweets and obesity and between healthy food, exercise and weight loss. When Shirley no longer fits through the barn door, her obesity stages an obstacle that has to be overcome. In a humorous, slapstick-like mode, the course of action is staged: for example when Shaun discovers numerous fattening sweets and pastries hidden in Shirley's thick fleece, or when he motivates her by placing delicious food in front of her as she runs on the treadmill. The association between food choices, exercise, weight and fitness is displayed in a direct manner. A second layer is added by recurring genre references and film quotations such as the genre of history drama and, more explicitly, the allusion to the cult film *Rocky* (John G. Avildsen, 1976) in the staging of the training programme leading to Shirley's successful transformation into an athlete, crowned by Shirley's ascension in slow motion onto a rubbish mountain. While these latter allusions can in fact only be grasped by a media-literate audience, the use of slapstick and of commenting music and sound, the style and outlook of the moulded plasticine clay figures and the modelled environment support the general sense of humour within the series.

Despite the clear narrative task of downsizing Shirley's weight, the food message in the episode is ambiguous. The food representations are used in a blatant, stereotypical manner, with sweets and pastries connoted as fattening and vegetables and exercises as markers of health and fitness. However, Shirley is back to her old self in the end, undermining the superficial message of a healthy diet. Children experience the course of action and the offered knowledge and information in media through the characters: it is via the characters that children tie in with the story. This opens a space for interpretation, since narratives usually allow for the adoption of multiple perspectives (Töpper and Prommer, 2004, p. 28). The example negotiates aspects of everyday life that young children can presumably relate to because food and health issues are usually part of family life. The overall topic of healthy (fruits and vegetables) and unhealthy (sweets and pastries) food thus offers points of attachment to which children can connect with their own experience. Yet the narration and the appeal of the episode work on more than one level: the second layer is the resolution of the obstacle (transforming Shirley), which might add an entertaining and pleasurable element for the young viewers. Shaun and his flock solve a problem – they perform agency – and with this the children are invited to experience the positive problem solving in an entertaining, humorous way. The lesson learned about food and obesity might thus be second-rank – according to the viewers' individual and socio-cultural background, the nutrition aspect can be of more or less relevance. In fact, Shirley's obesity is not explicitly judged: in the end, it is not framed as a problem that she is back to her old size. This adds to the possible interpretation space: the text determines not the function of food representation, but the practices of the recipients (Couldry, 2004, p. 129).

Teen lifestyle and distinction strategies in *Heroes of Olympus*

A second example of food to examine in the area of children's food media experiences is offered in the novel *The Heroes of Olympus: The Mark of Athena* (Rick Riordan, 2012). This fantasy novel, aimed at a pre-teen and teen audience, was one of the bestselling youth books in Germany in 2013. As a follow-up to the successful *Percy Jackson* series, the book combines adventurous action with romantic aspects. The story revolves around a group of demigods: Percy, Annabeth, Jason, Piper, Leo, Frank Zhang and Hazel. While the framing story displays the adventure of the group in a flying war-vessel towards Greece where their mission is to close the 'doors of death,' each character has her own stories and problems concerning self-esteem, love and friendship. In the end the 'doors of death' are indeed closed by the group, and all uncertainties and problems are solved.

Representations of food are heavily employed in the novel. Twenty-six direct references to food were identified, while the similar popular youth novels *Skulduggery Pleasant: Kingdom of The Wicked* (Derek Landy, 2012) and *Warrior Cats: Sunrise* (Erin Hunter, German release 2013) displayed fewer than ten relevant units of analysis. Most of the food references comprise an embellishing and atmospheric function. Only in five cases does food representation contain a narrative or dramaturgic function – either as magic devices helping the protagonists to survive their adventure, or diverted from their original use into a weapon. Yet the kind of food displayed is highly significant. Without exception, only fast food and snacks are featured during the story. Both protagonists and other-world inhabitants eat pizzas, burritos, bagels, quesadillas with avocado, cake, tea, doughnuts, cheeseburgers, marmalade, brownies, pastries, sandwiches, tortillas and 'Wonderbread.' They drink hot chocolate, Coca Cola, water or tea, displaying a globalised youth lifestyle. Food representations thereby not only embellish a scene, they also serve to characterise the protagonists. Pizza and Coca Cola, to take one example, are linked to the male characters, while the female protagonists eat bagels and drink tea and water.

Due to their frequency, their distribution over the whole course of the narrative and their importance within the scenes, food representations have a dominant status within the novel. But more than this they are part of a structuring pattern of the whole narration that divides the various storylines into action sequences, social sequences (the group and individual friendships) and romantic storylines (the heterosexual love stories into which all group members are drawn). While food is rarely used as a prop with a narrative or dramaturgical function during the action sequences, it is an important signifier for the social sequences of the book. Food embeds the fantastic stories of the seven demigods within the actual lives of preteens and teens, with fast food and snack food offering a point of contact and link to the lifeworlds of young people. At the same time, the specific kind of food ministers to the distinguishing strategies of adolescents towards their parents and towards the adult world.

Audience ratings: value and limitation

Charlton and Neumann (1988) emphasise the situational and cultural context of media reception without losing sight of the interactional perspective: the structural premises of the act of reception are accessible via the 'action-guiding topics' (Charlton and Neumann, 1988, p. 21). As Ralph Weiss (2000, 2001) elaborates, recurring on Bourdieu, these 'action-guiding topics' can be thought of as emerging from the 'practical sense' of media reception. These 'topics' can be small entities (such as hunger), but also more comprehensive topics such as becoming independent of one's parents. According to Weiss, such action-leading topics reflect back on the orientation pattern within the life-world (Weiss, 2000, p. 50), but also on the processes of media reception and appropriation. They are thus the key to the meaningful 'doing media' of the individual. In the analysis of media texts, the action-orienting topics are the key to understand the appeal of a specific media product (Bachmair, 1996; Mikos, 2001, p. 89). In both chosen examples, food can be considered to be a means by which the text addresses 'action-guiding topics' for the young audiences, while at the same time hinting at the limitation of this study: an audience study is needed as the missing link between the textual analysis and the content analysis. While this will be the necessary subsequent step in exploring the significance of media food for children in greater depth, the design of this study allows some inferences to be drawn between the collected media offerings and its potential audience on the basis of range of coverage.

As described, the sample consists of elements that can be considered most popular among the audience of 6- to 12-year-old children. Especially in regard to television, the information concerning the media audience is quite detailed. For other media, such as children's journals, less detailed information, such as sales figures or consumer surveys, was available. Relating this audience information to the actual sample allows a preliminary insight into the media preferences and media 'doings' of children.

Determining factors were found on the level of nationality/regionality on the level of age and on the level of gender. Most strikingly: Germany – at least for children up to 12 – is still a television nation. Watching television is the second most frequent leisure-time activity after learning and doing homework, and even more important than meeting with friends. Seventy-nine per cent of all children watch television on a daily basis: 97 per cent watch at least several times a week. Moreover, television is the medium from which children would prefer to abstain least of all media (KIM-Studie, 2012, pp. 10–12). Six- to nine-year-old children in Germany spend an average of 94 minutes in front of the television; 10- to

13-year-old children watch television 99 minutes per day (Feierabend and Klingler, 2013, p. 191). In contrast to other European countries such as Denmark, in Germany the internet has not yet gained dominance over television within the considered target group. In 2012 only 36 per cent of 6- to 13-year-old children in Germany accessed the internet on a daily basis, which makes them late adapters of the internet in comparison to many other European countries (Mascheroni and Ólafsson, 2013, p. 11). Secondly, within the sample, gender-related differentiations can be identified with regard to preferred media and preferred media products: video games in Germany are still a boys' medium, with boys playing more frequently than girls (KIM-Studie, 2012, p. 46), while books are a girls' medium: 58 per cent of girls are regular readers, compared to 39 per cent of boys (KIM-Studie, 2012, p. 25). Girls and boys also name different titles as their favourite across different media (KidsVA, 2012; KIM-Studie, 2012; Trend Tracking Kids®, 2012). Thirdly, age is a strong factor in media preferences. For example, SuperRTL is very popular among 6- to 9-year-old girls and boys. KiKa is especially liked by younger girls, but also has good shares within the other age groups apart from the older boys, while Nickelodeon is more popular with young boys. RTL, the channel with the biggest market share, has a big audience among older girls. Also pro7 is a channel that is preferred by older children, and here particularly boys.

Conclusion

Children in mid-childhood ages are not a homogeneous audience. In this study, investigation of the sample on the basis of range and preference emphasised certain particularities of media use with regard to nationality/regionality, age and gender. However, while a close look at detailed audience ratings can bring into focus an agglomeration of media practices and 'doings' that children perform with popular media formats and content, the structure and manifestation of these practices cannot be accessed via audience ratings or a textual analysis: an audience study is also necessary.

This contribution has provided a map of media food across the media actually used by children. In so doing, it has provided insights into the modality of food representations across a range of media. The study aimed to map out where children encounter food in the media they use, and how these food representations are staged and contextually embedded. It was demonstrated that children in Germany encounter food representations on a variety of different media, but that television is their most important access point to media food. When food is represented in media, sweets are often depicted, followed by beverages. Food representations are often embedded in a fictional format, such as an animation or a novel, addressing the audience with a narrative mode of appellation. While food is frequently staged as one embellishing aspect among several, in one-third of the representations the focus on food was evident. Remarkably, food representations on the internet played only a minor role in the study, though much food-related content is potentially available here.

Beyond the content analysis, this contribution has analysed the texts of media food, discussing the potential functions of food as anchored and favoured within the text with regard to its significance for children. The two examples presented here show that food representations can provide the potential for strategies of orientation and gratifications of agency, as well as lifestyle and distinction strategies. The content analysis, in conjunction with the textual analysis, has thus proved fruitful in building a foundation upon which further investigation and audience research into children's food-related media practices can now take place.

Notes

1 The chapter draws on data and findings from the research project 'Landkarte der medialen Lebensmittelwelt für Kinder,' commissioned by *KErn* – Kompetenzzentrum für Ernährung (November 2013 to June 2014). The project was conducted by the University of Bayreuth under the responsibility of Professor Dr Claas Christian Germelmann, Professor Dr Matthias Christen and Professor Dr Martin Huber, with Dr Susanne Eichner (lead researcher) and Johanna Held (MA) and Nicole Molitor (BA) (project fellows).
2 The education–entertainment approach focuses on the potential positive educational effects of media content. The approach has been elaborated by Singhal and Rogers (1999) and is based on Bandura's social learning theory. Accordingly, attitudes and behaviour are adopted by observance. Thus a media product that displays the idea of a positive protagonist who is rewarded, a negative protagonist who is punished and a transitional character who develops an acquired behaviour, is considered to be suitable and convincing to change the attitude and behaviour of a media audience. *Sesame Street* can be considered as the most prominent example within the tradition of education–entertainment; lately many 'serious games,' such as *Re-Mission* (HopeLap, 2006), a game for young cancer patients, or the political game *Frontiers* (by Tobias Hammerle, Georg Hobmeier, Sonja Prlic and Karl Zechenter, 2011) have adopted this strategy.
3 IZI (Internationales Zentralinstitut für das Jugend- und Bildungsfernsehen) is a research institute of the BR (PBS in Bavaria), ZDF and the State Media Authority of Bavaria (Bayrische Landeszentrale für neue Medien).
4 The *Cut the Rope* app was sold more than 100 million times for Android and iPhone. It received several awards, such as the BAFTA Award 2011 for 'Games/Handheld,' the Game Developers Choice Award 2011 'Best Handheld Game,' and the Pocket Gamer Award 2012 'Best Casual/Puzzle Game.'
5 The KIM Studie is a long-lasting research project of the MPFS (Medienpädagogischer Forschungsverbund Südwest). Since 1999 the representative study (c. 1,200 children and their mothers, personal interview, CAPI) is conducted on a regular two-year basis. The focus of the study is general media use, leisure-time activities, thematic interests, available media facilities and media commitment. The KIM Studie 2014 was not released at the time of going to press.
6 The latter category – flyers and brochures – was added for the special purpose of the *KErn*. Flyers and brochures were considered to be a potential way to apply a possible communication strategy. Purely auditive media (radio, music, audio drama) were not included for practical reasons.
7 In a third step, research was targeted within each medium to cover media products with intentional and educational food references. This data was labelled 'serious media' and was interpreted separately from the quota sample. While important for the aim of the overall study, the results of this third part of the study are not considered in this chapter.
8 A unit of analysis is defined as a sequence or section joint by coherence of sense: for example, an action sequence in narrative formats.

References

Arendt, K., 2010. Kann Fernsehen zu besserem Essen verführen? Zur Wirksamkeit von Entertainment-Education-Maßnahmen für Kinder am Beispiel der Kinderserie *LazyTown*. *Televizion* 23(1), pp. 28–31.

Arendt, K., 2013. *Entertainment-Education für Kinder: Potenziale medialer Gesundheitsförderung im Bereich Ernährung*. Baden-Baden: Nomos.

Ashton, D., 2004. Food advertising and childhood obesity. *Journal of the Royal Society of Medicine* 97, pp. 51–52.

Baacke, D., Sander, U., and Vollbrecht, R., 1990. *Lebenswelten sind Medienwelten*. Opladen: Leske + Budrich.

Bachmair, B., 1996. *Fernsehkultur: Subjektivität in einer Welt bewegter Bilder.* Opladen: Westdeutscher Verlag.

Buckingham, D., 2011. *The Material Child: Growing Up in Consumer Culture*. Cambridge: Polity Press.

Cebulla-Jünger, E., 1995. Früher oder später kriegen wir euch alle. Kinder als Zielgruppe und Werbeträger. In: Deutsche Gesellschaft für Ernährung, eds. *Wechselwirkung zwischen Ernährung und kindlichem Verhalten*. Kernen: DEG, pp. 92–102.

Charlton, M., and Neumann, K., 1988. Massenkommunikation als Dialog. Zum aktuellen Diskussionsstand der handlungstheoretisch orientierten Rezeptionsforschung. *Communications* 14(3), pp. 7–38.

Charlton, M., and Neumann-Braun, K., 1990. *Medienrezeption und Identitätsbildung*. Tübingen: Narr.

Couldry, N., 2004. Theorizing media as a practice. *Social Semiotics* 14(2), pp. 115–132.

Diehl, J.M., 2005. Macht Werbung dick? Einfluss der Lebensmittelwerbung auf Kinder und Jugendliche. *Ernährungsumschau* 52(2), pp. 40–47.

Eichner, S., 2014. *Agency and Media Reception: Experiencing Video Games, Film, and Television*. Wiesbaden: Springer VS.

Feierabend, S., and Klingler, W., 2013. Was Kinder sehen. Eine Analyse der Fernsehnutzung Drei- bis 13-Jähriger 2012. *Media Perspektiven* 4/2013, pp. 190–201.

Fiske, J., 2009 [1987]. *Television Culture*. London: Routledge.

Gamble, M., and Cotunga, N., 1999. A quarter century of TV food advertising targeted at children. *American Journal of Health Behavior* 23(4), pp. 261–267.

Gerbner, G., 1998. Cultivation analysis: An overview. *Mass Communication and Society* 1(3–4), pp. 175–194.

Götz, M., Holler, A., and Unterstell, S., 2010. 'Viel Gemüse essen und viel Milch trinken.' Wie Kindern durch die Integration von gesunder Ernährung in Caillou Alltagswissen vermittelt wird. *Televizion* 23(1), pp. 32–36.

Hastings, G., Stead, M., McDermott, L., Forsyth, A., McKintosh, A.M., Rayner, M., Godfrey, C., Caraher, M. and Angus, K., 2003. *Review of Research on the Effects of Food Promotion to Children. Final Reports and Appendices. Prepared for Food Standards Agency.* Glasgow: Centre for Social Marketing.

Hengst, H., 2014. Kinderwelten im Wandel. In: A. Tillmann, S. Fleischer, and K.U. Hugger, eds. *Handbuch Kinder und Medien*. Wiesbaden: Springer VS, pp. 17–29.

Hofmann, O., 2012. Kids report 2011. *Televizion* 25(1), pp. 27–29.

Iser, W., 1970. Die Appellstrukturen im Text. Unbestimmtheit als Wirkungsbedingung literarischer Prosa. In: R. Warning, ed. 1994. *Rezeptionsästhetik: Theorie und Praxis*. München: Fink, pp. 228–252.

Iser, W., 1980 [1978]. *The Act of Reading: A Theory of Aesthetic Response*. London: Johns Hopkins University Press.

Jensen, K.B., 1999. Knowledge as received. In: J. Gripsrud, ed. *Television and Common Knowledge*. London: Routledge, pp. 125–135.

KidsVA, 2012. *KidsVerbraucherAnalyse (KidsVA) 2012*. Berlin: Egmont Ehapa Media GmbH. http://www.egmont-mediasolutions.de/

KIM-Studie, 2012. *Kinder + Medien. Computer + Internet: Basisuntersuchung zum Medienumgang 6- bis 13-Jähriger*. Aktualisierte Ausgabe. Stuttgart: Medienpädagogischer Forschungsverbund Südwest.

Krüger, U.M., 2009. Zwischen Spass und Anspruch: Kinderprogramme im deutschen Fernsehen. *Media Perspektiven* 8, pp. 413–431.

Lampert, C., 2007. *Gesundheitsförderung im Unterhaltungsformat: Wie Jugendliche gesundheitsbezogene Botschaften in fiktionalen Fernsehangeboten wahrnehmen und bewerten*. Baden-Baden: Nomos.

Lewis, M.K., and Hill, A.J., 1998. Food advertisement on British children's Television: A content analysis and experimental study with nine-year olds. *International Journal of Obesity* 22(3), pp. 206–214.

Lücke, S., 2007. *Ernährung im Fernsehen, Eine Kultivierungsstudie zur Darstellung und Wirkung*. Wiesbaden: Springer VS.

Mascheroni, G., and Ólafsson, K., 2013. *Mobile Internet Access and Use Among European Children: Initial Findings of the Net Children Go Mobile Project*. Milano: Educatt.

Mikos, L., 2001. *Fern-Sehen: Bausteine zu einer Rezeptionsästhetik des Fernsehens*. Berlin: Vistas.

Niesto, H., 2009. Digitale Medien, soziale Benachteiligung und soziale Distinktion. *Medienpädagogik: Zeitschrift für Theorie und Praxis*. http://www.medienpaed.com/Documents/medienpaed/17/niesyt00906.pdf.

Paus-Haase, I., and Hasebrink, U., 2001. Talkshows im Alltag von Jugendlichen: Zusammenfassung der 'Talkshow-Studie.' In: U. Göttlich, F. Krotz, and I. Paus-Haase, eds. *Daily Soaps und Daily Talks im Alltag von Jugendlichen*. Opladen: Leske + Budrich, pp. 137–154.

Reckwitz, A., 2002. Toward a theory of social practices. A development in culturalist theorizing. *European Journal of Social Theory* 5(2), pp. 243–263.

Rössler, P., Lücke, S., Linzmaier, V., Steinhilper, L., and Willhöft, C., 2006. *Ernährung im Fernsehen: Darstellung und Wirkung: Eine empirische Studie*. München: Verlag Reinhard Fischer.

Rossmann, C., and Hastall, M.R., 2013. Gesundheitskommunikation als Forschungsfeld der deutschsprachigen Kommunikationswissenschaft: Bestandsaufnahme und Ausblick. In: C. Rossmann and M.R. Hastall, eds. *Medien und Gesundheitskomunikation. Befunde, Entwicklungen, Herausforderungen*. Baden-Baden: Nomos, pp. 9–19.

Singhal, A. and Rogers, E.M., 1999. *Entertainment-Education*. London: Taylor and Francis.

Theunert, H., 2005. Medienkonvergenz–eine Herausforderung für die medienpädagogische Forschung. In: H. Kleber, ed. *Perspektiven der Medienpädagogik in Wissenschaft und Bildungspraxis*. München: kopaed, pp. 111–124.

Theunert, H., Pescher, R., Best, P. and Schorb, B. 1992. *Zwischen Vergnügen und Angst–Fernsehen im Alltag von Kindern: Eine Untersuchung zur Wahrnehmung und Verarbeitung von Fernsehinhalten durch Kinder aus unterschiedlichen soziokulturellen Milieus*. Hamburg: Vistas.

Töpper, C., and Prommer, E., 2004. Dramaturgie heißt: Räume schaffen. Erzählmodi in Lernsendungen. *Televizion* 17(1), pp. 27–28.

Trend Tracking Kids®, 2012. Iconkids & Youth, International Research GmbH.

Vogelgesang, W., 1991. *Jugendliche Video-Cliquen: Action- und Horrorvideos als Kristallisationspunkte einer neuen Fankultur.* Opladen: Westdeutscher Verlag.

Vollbrecht, R., 1998. Wie Kinder mit Werbung umgehen. Ergebnisse eines DFG-Forschungsprojekts. In: H. Dichanz, ed. *Handbuch Medien: Medienforschung*. Bonn: Bundeszentrale für politische Bildung, pp. 188–195.

Vollbrecht, R., 2002. *Jugendmedien.* Tübingen: Niemeyer.

Wagner, U., and Theunert, H., 2007. Konvergenzbezogene Medienaneignung in Kindheit und Jugend. *Medienpädagogik: Zeitschrift für Theorie und Praxis der Medienbildung. Vol. 14*, http://www.medienpaed.com/Documents/medienpaed/14/wagner_theunert0712.pdf.

Wegener, C., 2008. *Medien, Aneignung und Identität: 'Stars' im Alltag jugendlicher Fans.* Wiesbaden: Springer VS.

Weiss, R., 2000. "Praktischer Sinn", soziale Identität und Fern-Sehen. Ein Konzept für die Analyse der Einbettung kulturellen Handelns in die Alltagswelt. *Medien und Kommunikationswissenschaft* 1(1), pp. 42–62.

Weiss, R., 2001. *Fern-Sehen im Alltag: Zur Sozialpsychologie der Medienrezeption.* Wiesbaden: Westdeutscher Verlag.

11 Children cooking media food

Exploring media (food) literacy through experimental methods

Stinne Gunder Strøm Krogager

During a creative co-production in a focus group, three 9th grade girls produce an anti-McDonald's campaign called: '*I'm hatin' it*'. It is starring an overweight man pouring sugar on a piece of cardboard. As he takes a bite you hear the voice-over: '*Do you have a craving for cardboard with sugar for lunch? Surely not? Quit McDonald's*', and the original music jingle sounds but with the altered wording: '*I'm hatin' it*'.

In preparation of this campaign, the producers' (the three girls') frame of reference according to ongoing discussions of McDonald's food is displayed and a wide repertoire of mediated discourses and myths are triggered; these are then merged in a novel and ironic way that display the producers' knowledge of media discourses and content.

This chapter exemplifies how we can study and demonstrate media knowledge and media literacy (Buckingham, 2007) through experimental methods: Making media users produce their own media content on food by building on established and well-known formulas and genres opens up to a wide range of media discourses. These discourses are not in themselves interesting, but how the participants use, transform and thus negotiate the established discourses on food demonstrate a high degree of media knowledge and media literacy. The participants in the empirical study, which the chapter is based on, convert originals and make ironic parodies and pastiches creating heterotopian places (Foucault and Miskowiec, 1986), as for example the anti-McDonald's campaign above.

Media are embedded in our everyday lives and they play a substantial role in our everyday practices, discourses, negotiations and sense making. Hence, studying the intersection between media and food in the lives of 10–16-year-old Danes should take into account:

> the whole range of practices in which media consumption and media-related talk is embedded, including practices of avoiding or selecting out, media inputs (. . .) such practices may not be part of what normally we refer to by 'media culture', but as practices oriented to media they are hardly trivial.
>
> (Couldry, 2004, p. 4)

Media cannot be separated from society and culture (Couldry and Hepp, 2013), which calls for attention when studying media use and media literacy. The empirical study, which this chapter builds on, makes 9th graders 'avoid and select media inputs', as Couldry puts it, to display practices and knowledge surrounding the intersection between media and food.

The framework

The empirical work presented in the chapter is part of a large cross-disciplinary research project at the University of Aarhus, Cool Snacks.[1] The overall aim of Cool Snacks was to develop healthier snack products targeted at Danish children and adolescents and to embed these snack products into a media and marketing plan (Grunert et al., 2016).

This subproject aimed at uncovering media use and whether we can identify a relation between media and food in the everyday lives of 10–16-year-olds. The subproject consists of a quantitative and a qualitative study; however, this chapter focuses only on some of the creative and practice-oriented parts of the qualitative study.[2] The qualitative study consists of twelve focus groups and ten individual in-depth interviews. Two pilot interviews were conducted in private homes (Olsen and Povlsen, 2010), whereas the ten following focus groups and the ten individual in-depth interviews were conducted in a school setting. Four different Danish schools (two private schools and two state schools), located in a countryside area, in a suburban area and in the centre of a city (well over 300,000 inhabitants), participated in the study. The interviews were carried out in 4th grade (10–11-year-olds) and 9th grade (15–16-year-olds) classes.[3] The participants were recruited through their class teachers, and the only criterion was that the participants must be different, socially and academically. This criterion was chosen to attain maximum variation within the sample (Neergaard, 2001), and the criterion also guided the choice of schools (private + state and geographical diversity). The two pilot interviews indicated that boys and girls have different preferences regarding both media and food. Consequently, the majority of the focus groups were divided into girls and boys groups: four all-girls focus groups and four all-boys focus groups, besides two focus groups mixing both genders. These two focus groups were meant as *test groups* to elucidate what happens to the interaction within a group of peers when gender is secluded in settings like this.

In the focus groups, the participants made the creative media co-productions (Thomson, 2008) consisting of media campaigns on food. The participants were asked to choose a food product, item or brand from among a pile of different pictures of food. Afterwards the groups were split into two teams. One team was given the task of arguing against the chosen food item and the other team was to argue in favour of it. This resulted in a 'communicative battle' in the focus groups. The objective of the creative and communicative co-productions (media campaigns for and against food) was partly to shift focus from the conversation, thus diverting the participants' attention from the interviewers, and partly to supply an alternative understanding (to that of the conversation) of the role that media and food play in the everyday lives of the participants (Gauntlett, 2005; Bragg and Buckingham, 2008; Buckingham, 2009; Johansson et al., 2010).

Figure 11.1 The four snack bowls

Another element that was added to the focus groups was snacking. Four bowls of snacks were placed on the table: one with mini carrots and cherry tomatoes, one with raspberries and blueberries, one with cinnamon buns and one with wine gums, liquorice and chocolate. The intention of serving snacks was to get the participants to talk about food without interviewers having to ask. In other words, the snack bowls were used as a *prompting* technique (McCracken, 1988, p. 37).

The purpose of the research design was to prioritise practices (doings) through creative co-productions and prompts (the snacks). These doings were supposed to function as an equivalent to dialogue (sayings); hence, the study produced varied data on everyday food consumption and media use, and it displayed a great level of media knowledge about food, and it suggested a link between media and food that will be unfolded in the following.

Media discourses on food – Adolescents working communicatively with messages on food

There was no set outline or instruction to follow doing the media campaigns for and against food products. Each team was given a pencil and paper and some used this for writing down slogans and drawing mascots, others did not use it at all. During the preliminary selection process, many of the food items (pictures of different food articles) were discussed: pros and cons of choosing a sophisticated gourmet setup, a decadent chocolate cake, a picture of a milk carton, a bag of marshmallows, a McDonald's cheeseburger, cauliflower, tomatoes or any other product or item. This selection method offered a broad insight into issues such as the obvious nutrition facts of the items in question, e.g. the content of calcium in milk, fat in crisps and other evident qualities, such as ecology, that the participants referred to when discussing the different choices. These facts were instrumental in the first arguments and discourses that the participants brought up in their discussion. Arguing pro milk, some of the participants indirectly referred to a campaign in which the Danish national soccer team, which is sponsored by the largest dairy factory in Denmark, promotes milk as a promoter of healthy lifestyle (calcium providing strong bones). In their final decision as to which item or product to work with for the campaign, the participants chose the food (or beverage) product based on an assessment of whether they could argue both sides (pro and con). The two all-boys focus groups chose alcohol and an organically bred chicken. The two all-girls groups chose crisps and a cheeseburger.

The boys group working with alcohol chose a picture of an orange-flavoured Bacardi Breezer. It was obvious that they were familiar with the beverage, and the Bacardi Breezer became representative of different alcoholic beverages of this type (e.g. Smirnoff Ice, Cult Shaker and Mokaï; all bottled alcoholic beverages). In their initial discussion, unspecified but provocative advertising campaigns for Cult Shaker were mentioned briefly, and it was clear that these campaigns constituted a shared frame of reference.[4]

The three boys arguing against the Bacardi Breezer drew on a wide range of television programs on alcohol and drug abuse that had been broadcasted in the years preceding the interview.[5] Some of these programs were specifically addressed at a young audience, and some of the programs were used for teaching and informing older students in schools about the dangers of drinking and taking drugs. The boys in this group combined facts (you might get sick or dehydrate when you drink) and taboos of drug and alcohol abuse (it can tear people apart, drinking can cause loosing someone close to you) with humoristic arguments (the need to urinate) in an ironic campaign whose sender was *Mothers against Bacardi.*

The three boys arguing in favour of the Bacardi Breezer made a more traditional commercial ad, drawing on the conventional theme setting for promoting alcoholic beverages: friendship, lovers and partying, starring beautiful young people. The ad shows that drinking Bacardi Breezer provides cohesion, it helps you open up to new people and you have fun when drinking. In addition to these stereotypical situations of an ad for alcoholic beverages, the boys argued that it tastes good and you become plastered!

Another focus group of boys in the 9th grade chose a picture of an organically bred chicken. The pro-argumentation drew on traditional health communication: *Eat a healthy diet, live well, eat chicken.* These boys were focused on ecology and the low fat and high energy levels in chicken meat in their key arguments in the campaign. The sender of the campaign was The Danish Veterinary and Food Administration (the UK equivalent: Food Standards Agency), and besides ecology and fat, energy level, the campaign plays on various issues: Chicken meat is cheap, it is easy to cook, it tastes good, it helps make your life better, it is Danish bred and freshly slaughtered.

The team arguing against chicken had some trouble finding arguments, but they ended up with a partly ironic campaign with the sender WSPA (World Society for the Protection of Animals). The boys discussed numerous related issues while making the campaign, e.g. animal welfare; battery hens walking on top of each other; long distances of transportation with poor conditions regarding space; and the risk of salmonella poisoning if the meat is prepared in the wrong way. The name of the campaign was: *Better conditions for chickens,* and the slogan was: 'Also chickens have feelings'. Besides, the team designed a catchphrase, reusing the voice-over from a television commercial for a large Danish supermarket chain, Bilka: 'Who can? Chicken can!' (playing on the Danish wording, which is lost in translation).

Figure 11.2 Picture of a cheeseburger

One of the girls groups chose the picture above of a cheeseburger. There was no brand name or indication of the origin of the burger, but the girls had no doubts that it was a McDonald's cheeseburger, and the picture prompted them to discuss the American documentary *Super Size Me* by Morgan Spurlock as well as several YouTube videos, e.g. testing how long it takes McDonald's foods to rot, and myths about chicken McNuggets being made of whole chickens (heads, feet and intestines included).

They created a very detailed anti-McDonald's campaign directed at young people. The purpose of the campaign was to call attention to the poor nutritious value of McDonald's food. The campaign consisted of an audio-visual television spot: An overweight man is pouring sugar on a piece of cardboard. He takes a bite as you hear the voice-over: '*Do you have a craving for cardboard with sugar for lunch? Surely not? Quit McDonald's.*' Next you hear the McDonald's jingle, but the original wording has been changed from: '*I'm lovin' it*' to '*I'm hatin' it*'. In addition to this television spot, the campaign encompasses posters located in urban spaces frequented by young people: The overweight man from the television spot reappears on the poster holding the piece of cardboard with sugar on it. As in the television spot, the writing on the poster says: '*Do you have a craving for cardboard with sugar for lunch?*'

The team of girls arguing in favour of the burger also worked with the burger as a representation of McDonald's. They made a commercial with the theme *family reunification* called '*Now the whole family is gathered*'. They played on the meaning of the term 'family reunification', using it to describe a family getting together or agreeing on going to McDonald's after a dispute about finances (the solution: Everyone can afford a family meal at McDonald's). The commercial reuses the

McDonald's brand as a family restaurant and tops it with a cost-effective argument.

The group of girls working with crisps did not design a regular campaign or commercial. Instead they listed some pro-arguments: satisfy hunger quickly, fast energy, a snack, cheap, sharable, easily accessible and delicious. They only listed a few con-arguments: high levels of fat, easily replaceable by healthier foods that are better for your body, bad conscience. These girls seemed uncomfortable about engaging in the creative work, and they finished the discussion quite early and moved on to talk about food and media outside the frames of the campaigns and commercials. Thus, it seemed that it was not the topic of food and media but rather the task that we asked them to undertake that they felt uncomfortable about. This group of girls was the only group not engaging fully in the creative co-productions. All other groups seemed amused about designing campaigns and commercials.

Transforming discourses: irony creating heterotopia

The co-productions showed some common structures in working with pro and con argumentation respectively. It appears that when arguing in favour of a food product (making commercials), to some extent, the participants in these focus groups use quite conventional settings and wordings. Often they mimic or mock well-known commercials, mixing factual argumentation with ironic paraphrases. Both as an international and a national Danish genre, commercials are very humoristic (Stigel, 2008), and in the co-productions made in the focus groups, this humoristic and ironic style is also used, e.g. by incorporating intertextuality and imitation drawing on the parody and pastiche genre (Hutcheon, 1994; Bruun, 2008). The short form of the commercial has developed its own system of expression, often using stereotypes and recognizable types of situations (Stigel, 2008), and the participants in the focus groups seem quite accustomed with these stereotypes as well as with the recognizable types of situations, e.g. young people partying and drinking Bacardi Breezer, and McDonald's as a family restaurant. The reason why the commercials are so effective in this co-productive setting is exactly their short form, the stereotypes and the recognizable types of situations. The short, reasonably fixed genre is easy to re-(and co-)produce because of the well-known originals; the recognisability of the structure and the different types of genre unleashes creativity and unconventionality within the established setting.

When arguing against the chosen food product (making campaigns), the participants were also familiar with the original form and specific campaign texts and visuals. Nevertheless, there seemed to be a bigger difference between the originals and the campaigns made in these focus groups. This might be due to the fact that although campaigns are a well-known genre, they might still not attract the same attention (from this target group) and be as frequently exposed as commercials, and to the fact that there is not always an original to mimic. The participants can almost always remember a commercial for the food products, the food categories or brands, whereas this is not always the case when working with the campaigns. Therefore some of the campaigns, e.g. the anti-McDonald's campaign and the

Mothers against Bacardi, twist the way of communicating and the role of the sender in more innovative (and again ironic) ways, as they do not always have an explicit original to mock or mimic. Some of the campaign-makers (focusing on ecologically bred chickens) use commercial slogans (from the supermarket chain Bilka), and thus the divide between the genres (commercial vs. campaign) is not rigid. This blurring of genres in some of the participants' co-productions might also reflect the blurring of genres in media-representations in general. Nevertheless, both commercials and campaigns are short forms communicating strategic messages according to relatively established formulas, and that makes them easily applicable to co-productions like this.

Co-producing food messages in well-known media genres such as the commercial and campaign genres opens up to a wide range of food-related issues that illustrates the participants' repertoire and knowledge about mediated discourses on food. Working with the co-production prompts, the participants discuss lifestyle, health, politics, environmental issues (e.g. pollution and ecology), animal welfare and risks without being directed to the discussions by the interviewers. This makes the co-productions effective in elucidating the not easily researchable intersections between media and food within the everyday lives of these 15–16-year-olds, and it is evident that they are knowledgeable about food-related issues communicated through the media. Furthermore, they are able to recreate food messages ironically and creatively, which shows that they are well versed in and familiar with both food discourses and the genres in which food messages occur. Using irony differently in the co-productions, the participants create heterotopias (Foucault and Miskowiec, 1986). They transform the established commercial and campaign discourses into discursive spaces that are incompatible and intricate by changing the settled discourses and thus creating *other* ambiguous spaces that resist the conventions they build on (Johnson, 2013). The hegemonic places of commercials and campaigns are eroded and new heterotopian spaces take form. This way the participants position themselves through irony, parody and pastiche.

Snacking in the focus groups – Sayings about doings

Another method used to prompt the participants in the focus groups (apart from doing co-productions) was serving snacks as described in *The Framework.* In all focus groups, four identical sets of snack bowls with healthy and unhealthy snacks were served, and across gender and age the participants talked about a fixed order in which you ought to eat the snacks. That is, the snacking practice was articulated; they *talked about how to do the snacking* (sayings about doings), and the order was absolute: At first, you have to eat the healthy snacks, then you can allow yourself to move on to less healthy bowls: carrots and tomatoes first, then berries, next cinnamon buns and finally sweets (Krogager, 2012). It was evident that all participants had been taught to eat healthy food before being rewarded with the unhealthy food. The unhealthy foods are the most desirable and used as a reward to our children when they have struggled through the healthy foods.

The precise origin of these practices is not evident from this particular study; nevertheless, it is obvious that they are shared norms among the participants, and this normative construction indicates certain discourses (and thus practices) in our mediatised society regarding food and eating (Hjarvard, 2008). Thus, when working with practices, discourses, negotiation and meaning, we have to take media into account, since they are so embedded in our society and everyday lives (Couldry, 2004). In this sense, norms should not be understood as authoritative rules accompanied by sanctions, but as practices through which suitable behaviour and performance are negotiated (Halkier and Jensen, 2008, p. 62). Practices always consist of both sayings and doings, i.e. practical activity and representations of this: 'A practice is thus a routinised way in which bodies are moved, objects are handled, subjects are treated, things are described and the world is understood' (Reckwitz in Warde, 2005, p. 135). The order in which you ought to eat healthy and unhealthy foods was a shared practice amongst the participants in this study, and they were prompted to articulate this routinised order by the snack bowls.

The upholding of other snacking-related practices and the matching performances were negotiated openly in some of the focus groups. In a girls' group, the participants kept a close eye on who ate what and how much. When one of the girls in this group was about to take the last cinnamon bun, another girl stopped her, trying to keep score of the cinnamon buns: How many buns had each participant had? The girl who tried to take the last bun had had several more than the other participants, and the group discussed this situation and agreed that the two girls who had had the least buns should share the last one. The girl who had tried to take the last bun (and who had already had the most) had broken the norm or challenged the practice regarding snacking in the focus group.

Another feature of the snack bowls that was not routinised and therefore posed an unfamiliar situation to the participants was the way in which the snacks were *not* introduced. When the snacks were placed on the table, the participants were not invited to sample them and the interviewers said nothing, and the groups managed this somewhat awkward situation very differently. Generally, the boys groups began snacking sooner than the girls groups, and in some of the girls groups, the interviewers eventually had to encourage the girls to sample the snacks – one group refrained from eating the snacks for almost an hour. Some of the girls groups asked politely if they could eat the snacks, most of the boys groups just started eating or made a joke – indirectly asking if they could eat the snacks. So the boys and the girls managed the snacks differently as regards how and when to start snacking, and thus it seemed that gender was performed differently in these focus groups when engaging in the social snacking practice (Butler, 1990, 1993). Seen from a more structuralist perspective, Roland Barthes describes food as 'a system of communication, a body of images, a protocol of usages, situations, and behavior' (Barthes, 1997, p. 20). Food functions as a system of communication in all cultures, because food-related practices are arranged in systems parallel with other cultural systems (Barthes, 1997; Krogager, 2012). Barthes uses the term 'grammar of foods' (Barthes, 1997, p. 22), and Mary Douglas uses the term 'code': 'If food is treated as a code, the message it encodes will be found in the pattern of social relations being expressed' (Douglas, 1997, p. 36). That is, food practices tell us something about social and gendered relations and structures; food is 'the link between social and bodily expressions of control, being both an aperture of the body and a social entrance and exit' (Bell and Valentine, 1997, p. 44), and this becomes apparent in these focus groups. The social and gendered *codes* and *grammars* are performed in practices that have to do with food and eating as well as representations of food.

Studying these situations and the associated know-how (the social organisation within the focus groups as well as what is displayed through the doings of the co-productions) as practices makes it feasible to deal with food and media in an everyday context. The practices constitute patterns that can be filled in different ways. The individual acts as the carrier of a practice, i.e. a carrier of patterns of bodily behaviour, routinised ways of understanding, know-how and desire. Media use and food consumption are embedded in our everyday practices and habits, and they are important constituents in the structuring of other routines in our everyday lives (Schatzki, 1996; Reckwitz, 2002; Shove, Trentman and Wilk, 2009). Hence, it can be difficult to separate the practices from each other, but working with doings like co-production and prompting like the snacks within social research settings may benefit the ability to distinguish and clarify the intersection between media use practices and food (consumption) practices. As described in the introduction, 'the whole range of practices' (Couldry, 2004, p. 4) surrounding media and food must be taken into consideration when studying the intersection between media and food and, thus, the effects of media-food messages. Taking this broad and practice-experimental approach to children's and adolescents' relation to media and food proves fruitful in the context of everyday life in which we can no longer separate media from society. Indeed, the approach has limitations too, which will be discussed in the following.

Doings displaying media-food literacy: gains and limitations

The practice-oriented approach used in this study benefits the study of knowledge about food and media as well as practices related to everyday media use and food consumption. When asked directly, participants in the study had difficulties in articulating where they come across food in the media, but doing the co-productions makes it clear which media and media texts the participants know and refer to in this specific social setting. The practices related to food and media intersect with other everyday practices, and this makes it difficult to identify and thus study these practices: We cannot ask media users and food consumers to report on their media-food consumption; we have to create spaces in which they can show it by their *doings*. Consequently, some of the ambiguous intersections between media and food are uncovered by studying which media and media texts the participants use and how they reprocess them.

The well-defined analytical distinction between doings and sayings presented here is to be understood as an analytical tool used for working methodologically with a prioritising of doings in empirical research: Of course, the empirical distinctions are not straightforward. Just as different practices intersect, the levels within a practice – that is, sayings and doings – overlap and blur. It is nevertheless obvious from the analysis of saying and doings that the participants in this study are familiar with a wide variety of media messages on food. They are able to reuse these messages (discursively and aesthetically) in new media-food productions by means of established mediated discourses and visual aesthetics in new and often ironic ways. Thus, the children and adolescents in this study appear to be well-informed media-food consumers and largely skilful media production creators.

However, there are some limitations to the specific practice-oriented methods used in the study. First and foremost, the pre-structuring of the methods should be addressed. Setting up a creative task within a focus group involves establishing a framework in which the participants *must* be creative. Making commercials and campaigns confines the participants to keep (more or less) within the communicative frame of arguing pro or con food items. Moreover, the communicative co-production was very loosely structured (see *The Framework*) to make room for creativity while working with a known and well-established genre. To some of the participants, this loose structure seemed to cause unease. On the other hand, a more confined structure might have reduced the creativity in other groups where originality thrived.

Besides, we have to take into account the school context in which these results were produced. The school constitutes a certain type of learning environment in which hierarchical structures and asymmetrical power relations are embedded. As researchers in the school setting, the interviewers may to some extent have substituted the teachers, and the participants may also to some degree have adopted their everyday roles of pupils trying to please the interviewers by demonstrating their knowledge and capability. In relation to some of the girls, this appeared to be the case in the beginning of the focus groups. Some girls were keen on displaying their skills and experience in regards to food, e.g. on different health issues or how to make homemade sushi. Yet, this demonstrating of knowledge lessened and vanished as the discussions took off. Regardless of the context, asymmetrical power relations can never be avoided in this type of qualitative research, in which focus is on knowledge production rather than knowledge collection; however, we have to be aware of the effect that the context may have on the outcome.

In addition, it is a basic understanding of the study to consider the *doings* and *sayings* and the negotiations and meanings embedded in these as elements of the social setting they are constructed within. Hence, in this context, media knowledge and food preference are the topics discussed by the participants in this specific social – single-gender – setting. The actual media use was also included in the study by means of quantitative database analysis, but these results have obviously not been included in the qualitative findings. In other words, the practices that were seen in these focus groups were the outcomes of a social construction; obviously a focus group, and thus what takes place in it, is not the same as everyday life. The interactions that unfold in these groups are influenced by the setting of the school and the power structure, which are embedded herein. Also, the existing relationship between the participants affects the way the participants act in this particular context where they are put together to discuss and work with media and food. Nonetheless, the practices that unfold in these focus groups are valid indications of some of the relations between media and food in the everyday lives of Danish children and adolescents.

The doings (co-production and prompting) in the focus groups functioned as a counterpart to dialogue (sayings). Practices regarding food and media and consisting of both sayings and doings were performed, and hence the study produced different yet interconnected data on everyday food and media practices and it proved a fruitful way to study knowledge of media, media literacy and media-food literacy in our mediatised society.

Notes

1 Funded by The Danish Agency for Science, Technology and Innovation: Food and Health (FØSU 2009–2012).
2 For further information on the quantitative part of the project, see Krogager (2012) and Krogager et al. (2015).
3 In this chapter, only the creative and practice-oriented sections of the focus groups in the 9th grade will be described. For further information on the individual in-depth interviews and the 4th grade groups, see Krogager (2012).
4 Cult is a Danish Company producing energy enhancing and alcoholic beverages. Different Cult Campaigns have repeatedly been discussed as chauvinistic and as out of step with proper advertising norms, and reported to The Consumer Council and the ombudsman. The campaigns often show naked bodies in sexual set-ups.
5 E.g. the two Danish documentary series: *Ultimatum* (Ultimatum), in which a small group of people confronting an addicted family member or friend demanding that the addicted person agrees to rehabilitation; and *100 dage uden stoffer* (A 100 days without drugs), in which a group of young people not willing to admit that they abuse alcohol or drugs are challenged to live 100 days without drinking/taking drugs.

References

Barthes, R., 1997. Towards a psychosociology of contemporary food consumption. In: Counihan, C. & Van Esterik, P., eds. *Food and culture: a reader*. New York: Routledge, pp. 28–35.

Bell, D. & G. Valentine, 1997. *Consuming geographies: we are where we eat*. London: Routledge.

Bragg, S. & D. Buckingham, 2008. 'Scrapbooks' as a resource in media research with young people. In: Thomson, P., ed. *Doing visual research with children and young people*. London: Routledge, pp. 114–131.

Bruun, H., 2008. The blue hippo in lifestyle television: on pastiche in television satire. *p.o.v. A Danish Journal of Film Studies*, 26(2), pp. 15–25.

Buckingham, D., 2007. *Media education: literacy, learning and contemporary culture*. Cambridge: Polity Press.

Buckingham, D., 2009. 'Creative' visual methods in media research: possibilities, problems and Proposals. *Media, Culture & Society*, 31(4), pp. 633–652.

Butler, J., 1990. *Gender Trouble: feminism and the subversion of identity*. New York: Routledge.

Butler, J., 1993. *Bodies that matter: on the discursive limits of "sex"*. New York: Routledge.

Couldry, N., 2004. Theorising media as practice. *Social Semiotics*, 14(2), pp. 115–132.
Couldry, N. & A. Hepp, 2013. Conceptualizing mediatization: contexts, traditions, arguments. *Communication Theory*, 23(3), pp. 191–315.
Douglas, M., 1997. Deciphering a meal. In: Counihan, C. & Van Esterik, P., eds. *Food and culture: a reader*. New York: Routledge.
Foucault, M. & J. Miskowiec, 1986. Of other spaces. *Diacritics*, 16(1), pp. 22–27.
Gauntlett, D., 2005. *Moving experiences: media effects and beyond*. Eastleigh, Australia: John Libbey & Co.
Grunert, K., S. Brock, K. Brunsø, T. Christensen, M. Edelenbos, H. Kastberg, S.G.S Krogager, L.H. Mielby, & K.K. Povlsen, 2016. Cool Snacks: a cross-disciplinary approach to healthier snacks for adolescents. *Trends in Food Science & Technology* 47, pp. 82–92.
Halkier, B. & I. Jensen, 2008. Det sociale som performativitet: et praksisteoretisk perspektiv p. analyse og metode. *Dansk sociologi*, 19(3), pp. 49–68.
Hjarvard, S., 2008. *Er verden af medier.* København: Samfundslitteratur.
Hutcheon, L., 1994. *Irony's edge: the theory and politics of irony*. London: Routledge.
Johansson, B., H. Brembeck, E. Ossiansson, K. Bergstrøm, L. Jonsson, H. Shanahan, & S. Hillén, 2010. Children as co-researchers – problems and possibilities. Paper presented at the 4th International Conference on Multidisciplinary Perspectives on Child and Teen Consumption. Interdisciplinarity, theory and practice. Linköpimg, Sweden, June.
Johnson, P., 2013. The geographies of heterotopia. *Geography Compass*, 7(11), pp. 790–803.
Krogager, S.G.S., 2012. When girls and boys use media and food. An experimental study of media use in 4th and 9th grade. Unpubl. PhD-dissertation. http://vbn.aau.dk/files/74878675/afhandling.pdf.
Krogager, S.G.S., K.K. Povlsen, & H.P. Degn, 2015. Patterns of media use and reflections on media: 10- to 16-year-old Danes. *Nordicom Review*, 36(2), pp. 97–112.
McCracken, G., 1988. *The long interview*. Newbury Park, CA: Sage Publications.
Neergaard, H., 2001. *Udvælgelse af cases i kvalitative undersøgelser*. Frederiksberg: Samfundslitteratur.
Olsen, S.G.S. & K.K. Povlsen, 2010. Eksperimenterende gruppeinterviews med børn om mad, medier og lyst. In: Bjørner, T., ed. *Den oplevede virkelighed: 11 eksempler på kvalitativ metode i praksis*. Aalborg: Aalborg Universitetsforlag, pp. 129–146.
Reckwitz, A., 2002. Toward a theory of social practices. *European Journal of Social Theory*, 5(2), pp. 243–263.
Schatzki, T.R., 1996. *Social practices: a Wittgensteinian approach to human activity and the social*. Cambridge: Cambridge University Press.
Shove, E., F. Trentman, & R. Wilk, 2009. *Time, consumption and everyday life: practice, materiality and culture*. New York: Berg.
Stigel, J., 2008. Humor I dansk tv-reklame. Et middel på tværs af livsstil? *Mediekultur*, 24(45), pp. 65–79.
Thomson, P., 2008. *Doing visual research with children and young people*. London: Routledge.
Warde, A., 2005. Consumption and theories of practice. *Journal of Consumer Culture*, 5(2), pp. 131–153.

12 Epilogue

Politics and the future of food and media

Kathleen LeBesco and Peter Naccarato

This collection has thoughtfully investigated cultural spaces where the meanings of foods and foodways are contested. In focusing on representations of food across media, the volume underscores the critical role such representations play in these debates as they shape how food and foodways evolve from the quotidian practice of eating into meaningful cultural signifiers. In doing so, they illuminate both historical and current trends in *how* food and foodways have served in two separate but crucially related ways, either enforcing a society's dominant ideologies or providing a means for resisting, challenging, and adapting them. In some chapters, authors interrogate food and foodways as agents of a neoliberal project of shaping individuals in ways that conform to their society's prevailing values and ideologies. At the same time, a number of the chapters have made use of Foucault's (1967) concept of heterotopia to investigate sites where resistance to such dominant ideologies flourishes. The question that we consider in the following pages is, as these spaces evolve, what shape the future of media food will take. We believe that taken collectively, the chapters in this volume point the way forward by eschewing the binary formulation of dominance/resistance in favor of analyses of hybrid food practices. At the same time, by extending their analyses from texts to practices, they also highlight the importance of shifting focus from how media food texts are created to how they are consumed.

We believe that while we have been witnessing the ascendance of all things foodie over the last two decades, we have now reached a point at which there is an equally compelling rejection of the values associated with foodie discourse and its oft-moralizing character. Yet there is no neat separation between the dominant power and the power that opposes it – they are flip sides of the same coin. We began to theorize this relationship in *Culinary Capital,* imagining the carnival, competitive eating, and junk food blogs as sites for resistance to the food orthodoxies espoused by the upper-middle class and their media counterparts. As we wrote there:

> what makes one path to culinary capital any more legitimate than another is completely arbitrary, but those with symbolic capital have the capacity to establish their route to culinary capital as privileged insofar as they are authorized to make their claims about reality stick.
>
> (Naccarato and LeBesco, 2012, p. 114)

Thus, the media representations that are at the center of this volume serve as crucial barometers for measuring current attitudes towards the role of food and foodways in promoting or challenging dominant ideologies.

Food and media in this volume

As several of the chapters in this volume make clear, media representations of food and foodways continue to endorse and circulate mainstream values in the service of neoliberalism. Caroline Nyvang's study of Danish cookbooks during the post-war period, for example, argues that they served as a space in which competing visions of the present and future were negotiated. From domestic consultants who 'would instruct and inform housewives and working mothers in matters related to nutrition, housekeeping and consumption' to the gourmands who emerged as culinary experts promoting food as an art form, Danish cookbooks in the postwar period became a site for negotiating individual and national identities. As such, Nyvang concludes, they underscore the fact that food can serve as 'a medium for the dissemination of wider political ideals' (Chapter 2).

This conclusion is reinforced by Joanne Hollows, whose analysis of the campaigning culinary documentary concludes that representations of working-class women's food practices function to taint their mothering skills and to suggest moral shortcomings. Framing her study as part of the broader investigation of 'how lifestyle and reality television formats have been used to naturalize neoliberal values and to draw distinctions between good and bad consumer-citizens' (Chapter 5), Hollows explores how representations of food and foodways utilize the related discourses of gender and class in projects of the self. In the particular example of the campaigning culinary documentary, middle-class values and ideologies are reinforced as they link working-class women's culinary inadequacies with failures of morality and citizenship.

Such policing of the lower class through privileged middle-class ideologies parallels a similar policing of masculinity through food as revealed by Fabio Parasecoli's analysis of the Weight Watchers' advertising campaign featuring former basketball player Charles Barkley. Focused on 'the role that cooking and eating play in performances of masculinity' (Chapter 6), Parasecoli examines how the Barkley advertising campaign exploits various tropes of masculinity to market Weight Watchers' diet plan to potential male clients. Mindful of traditional gendered values that dissociate masculinity from such an overt focus on body image while also acknowledging cultural shifts in such attitudes, this advertising campaign reveals the fluidity and adaptability of media food as it uses 'representations . . . about food-related practices and male body image [to] embrace and naturalize a set of expectations, values, and behaviors as constitutive elements of acceptable masculinity' (Chapter 6).

Like Nyvang, Hollows, and Parasecoli, several authors in this volume acknowledge that both historically and currently, media food function to circulate, promote, and reinforce mainstream values and ideologies, particularly those related to gender, class, ethnicity, and sexuality. Vera Alexander offers an historical perspective on this process and the ways in which it produces resistance in her analysis of how women negotiated traditional gender and class identities on the Canadian frontier. Through her analysis of food representation in prominent Canadian women's settlement narratives, Alexander reveals that in this context, food and foodways do not merely serve to reinforce normative ideologies of class and gender; instead, they play an important role among settlers seeking new identities and gender relations. In both their actual food practices and their writings about them, women settlers use foodways to reinvent their femininity and negotiate the transition 'from leisure class member to useful immigrant' (Chapter 3). Thus, the written representations of their frontier food practices serve as an important indication of the potential for food media to subvert as well as reinforce normative values and ideologies.

Katrine Meldgaard Kjær's chapter argues that Michelle Obama is a contemporary example of this process as she uses her food activism to reinforce dominant ideologies about family and femininity, while simultaneously undermining dominant ideologies about Black womanhood. Kjær concludes that by framing her food activism through the discourse of traditional motherhood and family, Michelle Obama creates and utilizes a 'third space to negotiate the delicate situation she is in as a politically active black First Lady' (Chapter 4). In doing so, she reveals the potential of media food to manipulate and disrupt the traditional values and ideologies that it often promotes.

Such negotiations are also evidenced in Jonatan Leer's examination of two popular Jamie Oliver television programs, *The Naked Chef* and *Jamie at Home,* and the Danish show *Spise med Price (Dining with Price),* hosted by brothers James and Adam Price. Arguing that contemporary representations of food and foodways can be read in relation to the so-called 'crisis of masculinity,' Leer concludes that these programs are indicative of 'homosocial heterotopias' through which competing discourses of masculinity can be negotiated. On the one hand, these programs reinforce several standard tropes of masculinity, particularly by distancing themselves from a more feminized narrative that links cooking to other family and domestic responsibilities. By 'disassociate[ing] cooking from a practical, everyday-oriented context' (Chapter 7), these shows carve out spaces in which cooking can be masculinized. Leer contends that they simultaneously create a countercultural space in which men not only escape the feminine, but can embrace and act out homosocial desires. In doing so, these programs do more than reinforce traditional definitions of masculinity; indeed, they 'create a mobile masculinity that lives up to the imperative of flexibility and negotiability of post-traditional masculinity' (Chapter 7).

Such mobility also comes into play as Habil Susanne Eichner, Stinne Gunder Strøm Krogager, Karen Klitgaard Povlsen, and Bente Halkier underscore the crucial role of the consumer in determining the extent and type of influence that media food has on both individuals and the broader cultural landscape. In her cross-media study from a German context, Eichner presents a mapping of where children encounter food in the media and which media representations the children prefer. She argues that 'food representations can provide the potential for strategies of orientation and gratifications of agency, as well as lifestyle and distinction strategies.' This argument, is also central to Krogager's study of the effects of media food messaging on children, reveals that her subjects do more than passively absorb information; rather, they are well equipped to transform and adapt such messaging, in some instances utilizing irony, parody, and pastiche as 'hegemonic places of commercials and campaigns are eroded and new heterotopian spaces take form' (Chapter 11). Similarly, Bente Halkier's study of motherhood food practices reveals the extent to which these practices, and the values that inform them, range from hegemonic and heteronormative to potentially heterotopic. Acknowledging that 'connections between food and motherhood are fraught with normative expectations' (Chapter 9), Halkier nonetheless concludes that as her subjects blend and adapt class- and ethnic-based culinary practices, they create 'hybrid and negotiated types of mothering through food [that] could be interpreted . . . as possibly heterotopian' (Chapter 9). Karen Klitgaard Povlsen's chapter demonstrates that media are neither monolithic nor overdeterminative of people's food preferences. People's use of multiple forms of media influences, and is influenced by, the 'taste regimes' of which they are a part, which also plays out in their unique foodways. Thus Povlsen, like Halkier and Krogager, reinforces the conclusion that it is not a matter of determining whether media food functions hegemonically to interpellate normative subjects or heterotopically to create spaces of liberation from such identities; rather, we need to acknowledge the adaptability and versatility of media food insofar as it can play both roles simultaneously.

The future of food and media

The future, it seems to us, will find people accruing culinary capital most successfully when they engage in foodways that hybridize the privileged and the resistant. And it is clear that such transitions are already afoot across the media food landscape as evidenced, for example, in the rapid shift away from instructional food programming. While this traditional type of food program may continue sporadically on not-for-profit venues like public television, such shows have all but disappeared on commercial television.

On the Food Network, for example, even instructional cooking shows that we have categorized as 'modern' insofar as they offer viewers 'a formula for bridging traditional expectations with the realities of contemporary life by acknowledging the limitations of time and money confronting many contemporary households' (Naccarato and LeBesco, 2012, p. 43) are increasingly marginalized. Proliferating on the Food Network and across the food media landscape are what we categorize as 'competitive' programs, where 'cooking is less about meeting one's familial or social obligations and more about confronting a challenge and persevering despite harsh conditions' (Naccarato and LeBesco, 2012, p. 44).

The rapid pace of this move away from instructional towards competitive programming signals a strange moment of opportunity as we witness an important shift in how media food negotiates the changing priorities and expectations of its consumers. At the same time, it underscores the fact that food – its procurement, preparation, and consumption – is not the primary consideration for media producers or consumers. Instead, as one of many platforms through which competitions can be waged, winners can be celebrated, and losers can be dismissed, food registers (bizarrely) less and less as a material object, and more as a lifestyle indicator. As such, it functions more as a floating signifier able to communicate a shifting and potentially more fluid set of values and ideologies than those directed towards a more passive consumer seeking to emulate the culinary and lifestyle skills of those deemed expert instructors. Of course, doubling down on competitive programming runs the risk of reinscribing traditional masculinity, power, and individualism at the expense of a connected, grounded food environment. But within the context of media, the shift away from instruction, toward decentralized power as consumers exercise greater influence on the types of food programming that is produced, holds promise.

And thus the future of food and media, we contend, will be one in which media producers continually adapt their programming in order to attract and maintain the interest of the viewing public. And for media consumers, it will be one in which they exercise increasing influence over the types of media produced and the extent to which that programming functions hegemonically or counter-hegemonically. This change has already begun to happen, as content produced by everyday citizens and circulated via channels like YouTube rivals the content produced by multinational media conglomerates and circulated via traditional network broadcasts, which has left the large media companies flummoxed about how to compete. Insofar as the priority of commercial media producers is maintaining ratings and profits, they will continue to respond to consumers who are becoming increasingly weary of a morally tinged foodie discourse and more proactive in shaping the evolving food media landscape.

As a result, food media will likely be less overtly instructional – both in terms of cooking techniques and modes of citizenship – and more focused on entertainment, which invites consumers to play a more active role in shaping what kinds of programs are produced, what values they promote, and how they are consumed by the viewing public. In other words, no longer content with being served a preset menu of food and food-related programming, consumers of media food are exercising more influence across the media food landscape. In doing so, they create a hybrid, potentially heterotopic space in which they intervene in what types of food programs are produced and the values and ideologies that such programming circulates. They are, in other words, 'consumer citizens looking for rewarding ways to remake their individual identities' by 'transform[ing] what constitutes privileged food discourse' (Naccarato and LeBesco, 2012, p. 115). It remains to be seen whether this transformation is insidiously neoliberal, or rather, empowering; more than likely, it is a combination.

References

Foucault, M.,1967. Des espaces autre. *Architecture, Mouvement, Continuité*, 5, pp. 46–49.
Naccarato, P. and K. LeBesco, 2012. *Culinary capital*. New York: Berg.

Index

For Product Safety Concerns and Information please contact our
EU representative GPSR@taylorandfrancis.com Taylor & Francis
Verlag GmbH, Kaufingerstraße 24, 80331 München, Germany